CONTEMPORARY MATHEMATICS

Titles in this Series

VOLUME 1 **Markov random fields and their applications**
Ross Kindermann and J. Laurie Snell

VOLUME 2 **Proceedings of the conference on integration, topology, and geometry in linear spaces**
William H. Graves, Editor

VOLUME 3 **The closed graph and P-closed graph properties in general topology**
T. R. Hamlett and L. L. Herrington

VOLUME 4 **Problems of elastic stability and vibrations**
Vadim Komkov, Editor

VOLUME 5 **Rational constructions of modules for simple Lie algebras**
George B. Seligman

VOLUME 6 **Umbral calculus and Hopf algebras**
Robert Morris, Editor

VOLUME 7 **Complex contour integral representation of cardinal spline functions**
Walter Schempp

VOLUME 8 **Ordered fields and real algebraic geometry**
D. W. Dubois and T. Recio, Editors

VOLUME 9 **Papers in algebra, analysis and statistics**
R. Lidl, Editor

VOLUME 10 **Operator algebras and K-theory**
Ronald G. Douglas and Claude Schochet, Editors

VOLUME 11 **Plane ellipticity and related problems**
Robert P. Gilbert, Editor

VOLUME 12 **Symposium on algebraic topology in honor of José Adem**
Samuel Gitler, Editor

Titles in this series

VOLUME 13 **Algebraists' homage: Papers in ring theory and related topics** Edited by S. A. Amitsur, D. J. Saltman and G. B. Seligman

VOLUME 14 **Lectures on Nielsen fixed point theory** Boju Jiang

VOLUME 15 **Advanced analytic number theory. Part I: Ramification theoretic methods** Carlos J. Moreno

VOLUME 16 **Complex representations of GL(2, K) for finite fields K** Ilya Piatetski-Shapiro

CONTEMPORARY MATHEMATICS

Volume 16

Complex Representations of GL(2, K) for Finite Fields K

Ilya Piatetski-Shapiro

AMERICAN MATHEMATICAL SOCIETY

Providence · Rhode Island

1980 *Mathematics Subject Classification.* Primary 20G05.

Library of Congress Cataloging in Publication Data
Piatetski-Shapiro, Ilya, 1929–
Complex representations of $GL(2, K)$ for finite fields K.
(Contemporary mathematics, ISSN 0271-4132; v. 16)
Includes bibliographical references and index.
1. Linear algebraic groups. 2. Representations of groups. 3. Modular fields. I. Title.
II. Series: Contemporary mathematics (American Mathematical Society); v. 16.
QA171.P47 1983 512′.2 82-24484
ISBN 0-8218-5019-9

Contents

FOREWORD

These are Lecture Notes of a course that I gave in Tel-Aviv University. The aim of these Notes is to present the theory of representations of $GL(2,K)$ where K is a finite field. However, the presentation of the material has in mind the theory of infinite dimensional representations of $GL(2,K)$ for local fields K.

I am very grateful to Moshe Jarden who took these Notes and worked them out. Without him it would have been completely impossible to prepare them.

This course and its Notes are the first outcome of the Cissie & Aaron Beare Chair in Algebra and Number Theory.

Ilya Piatetski-Shapiro
Tel-Aviv
November, 1982

Introduction

The aim of these notes is to give a description of the complex irreducible representations of the group $G = GL(2,K)$, where K is a finite field with $q > 2$ elements. In addition these notes should also serve as a motive for the study of the representation of $GL(2,K)$, where K is a local field. Therefore an attempt has been made to reprove theorems by not explicitly using the finiteness of K.

A central role in the description of the representations of G is played by the Borel subgroup consisting of all the matrices

$$b = \begin{pmatrix} \alpha & \beta \\ 0 & \delta \end{pmatrix} \qquad \alpha,\delta \in K^\times , \qquad \beta \in K .$$

If μ_1,μ_2 are characters of $K^\times$, then a character μ of B can be defined by $\mu(b) = \mu_1(\alpha)\mu_2(\delta)$. Let $\hat{\mu} = \mathrm{Ind}_B^G\mu$ be the induced representation. If $\mu_1 = \mu_2$, then $\hat{\mu}$ splits as the direct sum of a one-dimensional representation $\rho'_{(\mu_1,\mu_1)}$ which is given by formula $\rho'_{(\mu_1,\mu_1)}(g) = \mu_1(\det g)$, and a q-dimensional irreducible representation $\rho_{(\mu_1,\mu_1)}$. There are $q-1$ representations of each kind. If $\mu_1 \neq \mu_2$, then $\hat{\mu} = \rho_{(\mu_1,\mu_2)}$ is an irreducible representation of dimension $q+1$. There are $\frac{1}{2}(q-1)(q-2)$ representations of this kind. Irreducible representations that are not of the above types are of dimension $q-1$ and are called cuspidal representations. They are however also connected with linear characters in the following way. Let L be the unique quadratic extension of K and let ν be a character of $L^\times$ for which there does not exist a character χ of $K^\times$ such that $\chi(N_{L/K}z) = \nu(z)$ for every $z \in L^\times$. Such a ν is said to be non-decomposable. For each non-decomposable character ν of $L^\times$ we explicitly construct an irreducible representation ρ_ν of G and prove that it is cuspidal. Conversely, we prove that every cuspidal representation of G is of the form ρ_ν for some non-decomposable character ν of $L^\times$. Thus there are $\frac{1}{2}(q^2-q)$ cuspidal representations.

The connection between the irreducible representations of G and the characters of $K^\times$ and $L^\times$ gives rise to a reciprocity law. Let $W(L/K) = L^\times \cdot G(L/K)$ be the semi-direct product of $L^\times$ by $G(L/K)$. The irreducible representations of $W(L/K)$ (which is called the small Weil group) of dimension ≤ 2. The announced reciprocity law is a natural bijection between the two-dimensional representations of $W(L/K)$ (including the reducible ones) and the irreducible representations of G of dimension > 1.

Next we attempt to give explicit models for the irreducible representations of G. Let ψ be a non-unit character of K^+. The additive group K^+ can be canonically identified with the subgroup U of G consisting of all the matrices of the form

$$\begin{pmatrix} 1 & \beta \\ 0 & 1 \end{pmatrix} \qquad \beta \in K .$$

Therefore ψ can be also considered as a character of U. We prove that $\mathrm{Ind}_U^G \psi$ splits into the direct sum of all irreducible representations ρ of G of dimension > 1; each ρ appears with multiplicity 1. The space V_ρ on which ρ acts can therefore be embedded into $\mathrm{Ind}_U^G V_\psi$. Thus to each $v \in V_\rho$ there corresponds a function $W_v\colon G \to \mathbb{C}$ such that $W_v(ug) = \psi(u)W_v'(g)$ for every $u \in U$ and $g \in G$. The action of ρ on these functions is given by $W_{\rho(s)v}(g) = W_v(gs)$. The collection of all the W_v is called a Whittaker model for ρ. It has the following property: For all characters ω of $K^\times$ except possibly two there exists complex numbers $\Gamma_\rho(\omega)$ such that

$$(1) \qquad \Gamma_\rho(\omega) \sum_{x \in K^\times} W_v \begin{pmatrix} x & 0 \\ 0 & 1 \end{pmatrix} \omega(x) = \sum_{x \in K^\times} W_v \begin{pmatrix} 0 & 1 \\ x & 0 \end{pmatrix} \omega(x)$$

for every $v \in V_\rho$. If ρ is a cuspidal representation, then $\Gamma_\rho(\omega)$ is defined for every ω.

Among the Whittaker functions for ρ there is a special one, J_ρ, called the Bessel function of ρ, that satisfies

$$J_\rho(gu) = J_\rho(ug) = \psi(u)J_\rho(g) \quad \text{for } u \in U, \; g \in G .$$

Further, $J_\rho(1) = 1$ and $J_\rho(u) = 0$ for $u \in U$ and $u \neq 1$. Substituting this function for W_v in (1) we have

$$\Gamma_\rho(\omega) = \sum_{x \in K^\times} J_P \begin{pmatrix} 0 & 1 \\ x & 0 \end{pmatrix} \omega(x) .$$

This formula is then used in order to express $\Gamma_\rho(\omega)$ in terms of Gauss sums: If $\rho = \rho_{(\mu_1,\mu_2)}$ is a non-cuspidal representation of G, then

$$\Gamma_\rho(\omega) = \frac{\omega(-1)}{q} G_K(\mu_1^{-1}\omega^{-1},\psi)G_K(\mu_2^{-1}\omega^{-1},\psi) .$$

If $\rho = \rho_\nu$ is a cuspidal representation, then

$$\Gamma_\rho(\omega) = \frac{\nu(-1)}{q} G_L(\nu\cdot(\omega\circ N_{L/K})^{-1}, \; \psi\circ Tr_{L/K}) .$$

The Gauss sum $G_K(\chi,\psi)$ is defined for a character ψ of $K^\times$ and a character ψ of K^+ by

$$G(\chi,\psi) = \sum_{x \in K^\times} \chi(x)\psi(x) .$$

In particular it follows that in every case $|\Gamma_\rho(\omega)| = 1$.

All these results are finally applied in order to compute the characters table for G.

Chapter 1. Preliminaries: Representation theory; the general linear group

In the first three sections of this chapter we bring all the definitions and theorems about linear representations of finite groups that we need in these notes. We refer to Serre [2] and to Lang [1] for the proofs. The remaining two sections are devoted to a description of the group-theoretical properties of $GL(2,K)$, where K is a finite field.

1. Linear representations of finite groups.

Let V be a finite dimensional vector space over the field $\mathbb{C}$ of the complex numbers. Denote by $Aut(V)$ the group of all automorphisms of V. Let G be a finite group. A _linear representation_ of G in V is a homomorphism ρ of G into $Aut(V)$. V is said to be the _representation space_ of ρ and is also denoted by V_ρ. We shall also say that G _acts on_ V_ρ _through_ ρ. The _dimension_ of ρ is defined to be the dimension of V_ρ and is denoted by $\dim \rho$. Two representations ρ and ρ' of G are said to be _isomorphic_, if there exists an isomorphism $\theta\colon V_\rho \to V_{\rho'}$ such that $\theta\circ\rho(g) = \rho'(g)\circ\theta$ for every $g \in G$. We shall usually identify isomorphic representations.

A representation of G of dimension 1 is a homomorphism μ of G into the multiplicative group $\mathbb{C}^\times$ of $\mathbb{C}$. Such a representation is called in these notes a _character_ of G. In particular, the unit character is the homomorphism of G into $\mathbb{C}^\times$ obtaining the value 1 for every $g \in G$.

Let ρ be a representation of G and let H be a subgroup of G. Suppose that μ is a character of H for which there exists a non-zero $v \in V_\rho$ such that $\rho(h)v = \mu(h)v$ for every $h \in H$. Then μ is said to be an eigenvalue of H (with respect to ρ) and v is said to be an eigenvector of H that belongs to μ.

Again consider a representation ρ of G and let V' be a subspace of $V = V_\rho$ which is left invariant by $\rho(g)$ for every $g \in G$. In this case we say that V' is left invariant by G or that V' is a G-subspace of V. Then the restriction map of $\rho(g)$ to V' gives rise to a representation ρ' of G with V' as its representation space. This representation is said to be a sub-representation of ρ and we write $\rho' \leq \rho$. By a theorem of Maschke V' has a complement in V, i.e., there exists another G-subspace V'' of V such that $V = V' \oplus V''$ (c.f. Serre [2, p. 18]). Let ρ'' be the corresponding subrepresentation of ρ. Then ρ is said to be a direct sum of ρ' and ρ'' and we write $\rho = \rho' \oplus \rho''$. Clearly $\dim \rho = \dim \rho' + \dim \rho''$. The direct sum of n representations of G, all isomorphic to ρ, is denoted by $n\rho$. A representation ρ of V is said to be irreducible if it does not have a sub-representation ρ' of a lower dimension. By the theorem of Mascke this is equivalent to saying that ρ cannot be decomposed as a direct sum $\rho = \rho' \oplus \rho''$ with $\dim \rho' < \dim \rho$. It follows that every representation ρ of G can be represented as a direct sum $\rho = \bigoplus_{i=1}^{k} n_i \rho_i$, where the ρ_i are distinct (i.e., non-isomorphic) irreducible representations of G. This decomposition of ρ is unique, up to the order of the summands (c.f., Serre [3, p. 34]).

There are only finitely many irreducible representations $\rho_1, \ldots, \rho_n$ of G. Their number h is equal to the number of the conjugacy classes of G (c.f., Serre [3, p. 32]). Their dimensions satisfy the formula

$$(1) \qquad \sum_{i=1}^{n} (\dim \rho_i)^2 = |G| .$$

If G is abelian, then (1) implies that the irreducible representations of G are of dimension 1 (i.e., they are characters) and that their number is equal to $|G|$, which is in this case the number of the conjugacy classes of G. Further, the set of characters of G forms a multiplicative group $\hat{G}$ which is isomorphic to G. If $1 \neq \chi \in \hat{G}$, then we have the following orthogonality relation $\sum_{g\in G} \chi(g) = 0$. A lemma of Artin says that the characters of G are linearly independent, i.e., if a_χ are complex numbers such that $\sum_{\chi\in\hat{G}} a_\chi \chi(g) = 0$ for every $q \in G$, then $a_\chi = 0$ for all $\chi \in G$ (cf. Lang [1, p. 209]). Now, G is canonically isomorphic to the dual $\hat{\hat{G}}$ of $\hat{G}$. Hence, the dual to this lemma is also true: If b_g are complex numbers such that $\sum_{g\in G} b_g \chi(g) = 0$ for every $\chi \in G$, then $b_g = 0$ for all $g \in G$.

If G is again an arbitrary finite group, then we deduce that it has $(G:G')$ characters, where G' is the commutator subgroup of G. Another consequence of formula (1) is that if distinct irreducible representations $\rho_1,\ldots,\rho_n$ of G satisfy $\sum_{i=1}^{n} (\dim \rho_i)^2 = G$, then they are all the representations of G.

Let ρ be a representation of a finite group G. Then V_ρ can be also considered as a module over the group-ring $\mathbb{C}[G]$. If ρ' is an additional representation of G, then we write $(\rho,\rho') = \dim \mathrm{Hom}_{C[G]}(V_\rho,V_{\rho'})$. The form (ρ,ρ') is clearly symmetric and bilinear with respect to direct sums. If both ρ and ρ' are irreducible, then, by a lemma of Schur, $(\rho,\rho') = 1$ if $\rho = \rho'$ and $(\rho,\rho') = 0$ if $\rho \neq \rho'$ (cf. [2, p. 25]). It follows that two arbitrary representations ρ and ρ' are <u>disjoint</u>, i.e., have no common irreducible subrepresentation, if and only if $(\rho,\rho') = 0$. In particular, an irreducible representation ρ <u>appears</u> in a representation ρ', i.e., $\rho \lesssim \rho'$, if and only if $(\rho,\rho') \neq 0$; indeed (ρ,ρ') is equal to the multiplicity in which ρ appears in ρ'.

Let $\mathrm{End}_{\mathbb{C}[G]}V_\rho = \mathrm{Hom}_{\mathbb{C}[G]}(V_\rho,V_\rho)$. It is an algebra over $\mathbb{C}$ called the <u>Schur algebra</u>. If ρ is irreducible, then $\mathrm{End}_{\mathbb{C}[G]}V_{n\rho}$ is isomorphic to $M_n(\mathbb{C})$, the algebra of all $n\times n$ matrices over $\mathbb{C}$. If $\rho = \oplus n_i\rho_i$ is the canonic decomposition of a representation ρ, then, by Schur's lemma, $\mathrm{End}_{\mathbb{C}[G]}V = \oplus M_{n_i}(\mathbb{C})$. Hence $(\rho,\rho) = \dim \mathrm{End}_{\mathbb{C}[G]}V_\rho = \sum n_i^2$. It follows that ρ has no multiple components, i.e., that $n_i = 1$ for all i, if and only if $\mathrm{End}_{\mathbb{C}[G]}V_\rho$ is commutative. In this case $\dim \mathrm{End}_{\mathbb{C}[G]}V_\rho$ is the number of components of ρ.

Finally consider a vector space V of dimension n over $\mathbb{C}$. Every base $v_1,\ldots,v_n$ of V canonically defines an isomorphism $\mathrm{Aut}\, V \cong GL(n,\mathbb{C})$ (= the group of all $n\times n$ invertible matrices over $\mathbb{C}$). If $\rho\colon G \to \mathrm{Aut}\, V$ is a representation of V, then we define $\chi_\rho(g)$ to be the trace of $\rho(g)$, where $\rho(g)$ is now considered as an element of $GL(n,\mathbb{C})$ via the above isomorphism. Clearly $\mathrm{tr}\, \rho(g)$ does not depend on the choice of the basis $v_1,\ldots,v_n$ of V. Hence $\chi_\rho\colon G \to \mathbb{C}$ is a well defined function, called the <u>character of</u> ρ. It is constant on conjugacy classes. Also $\chi_{\rho_1\oplus\rho_2} = \chi_{\rho_1} + \chi_{\rho_2}$. Therefore χ_ρ is said to be <u>irreducible</u> if ρ is irreducible. If $\dim \rho = 1$, then $\chi_\rho = \chi$. In general one defines $\dim \chi_\rho = \dim \rho$ and refers to χ_ρ as a <u>higher dimensional character</u>.

2. Induced representations.

Let G be a finite group and let H be a subgroup operating on a finite dimensional $\mathbb{C}$-vector space W through a representation $\tau\colon H \to \mathrm{Aut}\, W$. Define a vector space V to be the set of all functions $f\colon G \to W$ that satisfy

$$f(hg) = \tau(h)f(g) \quad \text{for all} \quad h \in H \quad \text{and} \quad g \in G\ .$$

Thus, in order to define an element $f \in V$, it suffices to give its values on a system of representatives H/G of the left classes of G modulo H. Define an operation of G on V by

$$(sf)(g) = f(gs) \quad \text{for} \quad s,g \in G \quad \text{and} \quad f \in V\ .$$

The $\mathbb{C}[G]$ - Module V thus obtained is called the <u>induced module of</u> W <u>from</u> H <u>to</u> G and is denoted by $\mathrm{Ind}_H^G\tau$.

We embed W in V by mapping each $w \in W$ onto the function $f_w\colon W \to \mathbb{C}$ defined by $f_w(g) = \tau(g)w$ if $g \in H$ and $f_w(g) = 0$ if $g \in G-H$. Clearly this is a $\mathbb{C}[H]$ - modules embedding. The image of W in V consists of all the functions $f \in V$ that vanish on $G-H$.

Let now $G = \dot{\bigcup}_{r \in R} rH$ be a decomposition of G into left classes modulo H. For every $f \in V$ and for every $r \in R$ we define a function $f_r \in V$ by $f_r(g) = f(g)$ if $g \in Hr^{-1}$ and $f_r(g) = 0$ otherwise. Then $r^{-1}f_r$ belongs to W (after identifying W with its image in V) and $f = \sum_{r \in R} r(r^{-1}f_r)$. Thus V is isomorphic to $\bigoplus_{r \in R} rW$. In particular we have that $\dim V = (G{:}H)\dim W$. Using this isomorphism one obtains also a canonical isomorphism $V \cong \mathbb{C}[G]\otimes_{\mathbb{C}[H]}W$, where G operates on the right-hand side by multiplication on the left of the first factor. This form of the induced representation is convenient to prove the following fundamental properties. (a) Transitivity: If J is a subgroup of H and $\tau\colon J \to \operatorname{Aut} U$ is a representation of J, then

$$\operatorname{Ind}_J^G U = \operatorname{Ind}_H^G(\operatorname{Ind}_J^H U) \ .$$

(b) Frobenius reciprocity theorem: With the above notation let E be a $\mathbb{C}[G]$-module and denote by $\operatorname{Res}_H^G E$ the $C[H]$-module obtained from E by considering only the action of H. Then we have the following canonical isomorphism:

$$\operatorname{Hom}_{\mathbb{C}[G]}(\operatorname{Ind}_H^G W, E) \cong \operatorname{Hom}_{\mathbb{C}[H]}(W, \operatorname{Res}_H^G E)$$

(cf. [3, p. 23]). In particular,

$$\dim \operatorname{Hom}_{\mathbb{C}[G]}(\operatorname{Ind}_H^G W, E) = \dim \operatorname{Hom}_{\mathbb{C}[H]}(W, \operatorname{Res}_H^G E) \ .$$

If τ and σ are the representations of H and G that correspond to W and E, respectively, then the last equality can be rewritten, in the notation of section 1, as

$$(\operatorname{Ind}_H^G \tau, \sigma)_G = (\tau, \operatorname{Res}_H^G \sigma)_H \ .$$

In particular, if both τ and σ are irreducible, then the multiplicity of σ in $\operatorname{Ind}_H^G \tau$ is equal to the multiplicity of τ in $\operatorname{Res}_H^G \sigma$.

Finally, if τ is a representation of a subgroup H of a group G and $\rho = \operatorname{Ind}_H^G \tau$, then χ_ρ can be calculated from χ_τ by the following formula

$$\chi_\rho(g) = \frac{1}{|H|}\sum_{r\in G}\tilde{\chi}_\tau(sgs^{-1}) = \sum_{r\in R}\tilde{\chi}_\tau(rgr^{-1}) ,$$

where $\tilde{\chi}_\tau$ is the function on G that vanishes outside H and coincides with χ_τ on H; R is a system of representatives of right classes of G modulo H (cf. [2, p. 72]).

3. The Schur algebra.

Proposition 3.1: Let H and J be subgroups of a finite group G. Let ρ and σ be representations of H and J, respectively. Then $\mathrm{Hom}_{\mathbb{C}[G]}(\mathrm{Ind}_H^G V_\rho, \mathrm{Ind}_J^G V_\sigma)$ is isomorphic to the vector space of all functions $F: G \to \mathrm{Hom}_{\mathbb{C}}(V_\rho, V_\sigma)$ satisfying

(1) $$F(jgh) = \sigma(j)\circ F(g)\circ\rho(h)$$

for all $j \in J$, $g \in G$ and $h \in H$.

Proof: Let $\hat{\rho} = \mathrm{Ind}_H^G\rho$, $\hat{\sigma} = \mathrm{Ind}_J^G\sigma$ and $n = (G:H)$. Denote by F' the vector space of all functions

$$\varphi: G\times G \to \mathrm{Hom}_{\mathbb{C}}(V_\rho, V_\sigma)$$

that satisfy

(2) $$\varphi(jg_1, hg_2) = \sigma(j)\circ\varphi(g_1,g_2)\circ\rho(h)^{-1}$$

for all $j \in J$, $h \in H$ and $g_1, g_2 \in G$. For every $\varphi \in F'$ we define an element $T_\varphi \in \mathrm{Hom}_{\mathbb{C}}(V_{\hat{\rho}}, V_{\hat{\sigma}})$ as follows: If $f \in V_\rho$, then $T_\varphi f: G \to V_\sigma$ is the map defined by

(3) $$(T_\varphi f)(g) = \frac{1}{n}\sum_{r\in G}\varphi(g,r)(f(r)) ;$$

clearly the map $\varphi \to T_\varphi$ is a homomorphism $F' \to \mathrm{Hom}_{\mathbb{C}}(V_{\hat{\rho}}, V_{\hat{\sigma}})$. It is injective. Indeed, suppose that $T_\varphi = 0$. Let $s \in G$, let $v \in V_\rho$ and define a function $f_{sv} \in V_\rho$ by

$$f_{sv}(g) \begin{cases} \rho(h)v & \text{if } g = hs \\ 0 & \text{if } g \notin Hs . \end{cases}$$

Then substituting $f = f_{sv}$ in (3) we have by (2) that $\varphi(g,s)v = 0$. Hence $\varphi(g,s) = 0$, i.e., $\varphi = 0$.

The dimension of F' is equal to $(G:H)(G:J)(\dim\rho)(\dim\sigma)$ by (2). This is also the dimension of $\mathrm{Hom}_{\mathbb{C}}(V_{\hat\rho},V_{\hat\sigma})$. Hence T is an isomorphism.

Denote now by F'_G the subspace of all $\varphi \in F'$ such that $T_\varphi \in \mathrm{Hom}_{\mathbb{C}[G]}(V_{\hat\rho},V_{\hat\sigma})$. Clearly $\varphi \in F'_G$ if and only if

$$(4) \qquad \sum_{r\in G} \varphi(g,rx^{-1})(f(r)) = \sum_{r\in G} \varphi(gx,r)(f(r))$$

for all $f \in V_{\hat\rho}$. Substituting $f = f_{sv}$ in (4), we have that (4) is equivalent to the condition

$$(5) \qquad \varphi(g,rx^{-1}) = \varphi(gs,r) \quad \text{for all } g,r,x \in G .$$

For every function $F\colon G \to \mathrm{Hom}_{\mathbb{C}}(V_\rho,V_\sigma)$ that satisfies (1), we define a function $\varphi\colon G\times G \to \mathrm{Hom}_C(V_\rho,V_\delta)$ by

$$(6) \qquad \varphi(g_1,g_2) = F(G_1g_2^{-1}) .$$

Then φ satisfies (5) and thus it belongs to F'_G. Conversely, starting from φ in F'_G, we define an $F\colon G \to \mathrm{Hom}_{\mathbb{C}}(V_\rho,V_\sigma)$ by

$$F(g) = \varphi(g,1) .$$

Then F satisfies (1) and the φ defined by (6) coincides with the one we started with. Thus F is isomorphic to F'_G.

For every $F \in F$ denote by T_F the element of $\mathrm{Hom}_{\mathbb{C}[G]}(V_{\hat\rho},V_{\hat\sigma})$ defined by

$$(7) \qquad (T_Ff)(g) = \frac{1}{n} \sum_{r\in G} F(gr^{-1})(f(r)) .$$

Then the map $F \to T_F$ is the desired isomorphism. //

<u>Corollary 3.2</u>: In the notation of Proposition 3.1 we have

$$(\mathrm{Ind}_H^G\rho,\ \mathrm{Ind}_J^G\sigma) \leq |J\backslash G/H|(\dim\rho)(\dim\sigma)$$

where $J\backslash G/H$ denotes the set of double classes of G modulo J and H.

The most interesting conclusion of Proposition 3.1 arises in the special case where $H = J$ and $\rho = \sigma$. In this case $\mathrm{Hom}_{\mathbb{C}[G]}(\mathrm{Ind}_H^G V_\rho,\ \mathrm{Ind}_J^G V_\sigma) = \mathrm{End}_{\mathbb{C}[G]}(V_{\hat{\rho}})$, the Schur algebra of $\hat{\rho}$. The bijection between F and this algebra established in Proposition 3.1 turns F into an algebra and the product between two elements F_1 and F_2 of F is given by

$$(8)\qquad (F_1 * F_2)(g) = \frac{1}{(G:H)} \sum_{s\in G} F_1(gs^{-1})F_2(s)\ .$$

This can be easily verified from the basic relation $T_{F_1}T_{F_2} = T_{F_1 * F_2}$ and the definition (7).

4. The group GL(2,K).

In this section we fix our notation for the rest of these notes.

Let K be a finite field with q elements and suppose that $q > 2$. We denote by G the group $GL(2,K)$ of all 2×2 invertible matrices with entries in K. We further reserve some letters for distinguished subgroups of G that will concern us in the sequel. The letter B stands for the <u>Borel</u> subgroup of G consisting of all upper triangular matrices

$$B = \left\{\begin{pmatrix}\alpha & \beta\\ 0 & \delta\end{pmatrix} \middle|\ \alpha,\delta \in K^\times;\quad \beta \in K\right\}.$$

Clearly $|B| = (q-1)^2q$. Straightforward calculations show that the matrix $w = \begin{pmatrix}0 & 1\\ 1 & 0\end{pmatrix}$, together with the matrices $\begin{pmatrix}1 & 0\\ \gamma & 1\end{pmatrix}$, $\gamma \in K$, form a system of representatives for the left (and also for the right) classes of G modulo B. Hence $(G:B) = q+1$ and thus $|G| = (q-1)^2q(q+1)$. The indempotent matrix w will play an important role in the sequel.

B is a solvable group. Indeed, B contains the normal abelian subgroup

$$U = \left\{\begin{pmatrix} 1 & \beta \\ 0 & 1 \end{pmatrix} \middle| \quad \beta \in K\right\}$$

of all unipotent upper triangular matrices. This group is isomorphic to the additive group K^+ of the field K. Indeed

$$\begin{pmatrix} 1 & \beta \\ 0 & 1 \end{pmatrix}\begin{pmatrix} 1 & \beta' \\ 0 & 1 \end{pmatrix} = \begin{pmatrix} 1 & \beta+\beta' \\ 0 & 1 \end{pmatrix} .$$

We shall therefore sometimes identify an element β of K with the corresponding matrix of U. The quotient group B/U is isomorphic to the <u>Cartan</u> group

$$D = \left\{\begin{pmatrix} \alpha & 0 \\ 0 & \delta \end{pmatrix} \middle| \quad \alpha,\delta \in K^\times\right\}$$

of all diagonal matrices. It is isomorphic to $K^\times \times K^\times$ and hence is abelian. Clearly $U \cap D = 1$ and $UD = B$. Hence B is the semi-direct product of U by D. Simple calculation shows that U is the commutator subgroup of B. (Here we are using the assumption $q > 2$. In the case $q = 2$ we have $B = U$ and $B' = 1$.) In particular it follows that B has exactly $(q-1)^2$ characters.

Another important normal subgroup of B is

$$P = \left\{\begin{pmatrix} \alpha & \beta \\ 0 & 1 \end{pmatrix} \middle| \quad \alpha \in K^\times, \quad \beta \in K\right\}$$

of order $(q-1)q$ and of index $q-1$ in B. The center

$$Z = \left\{\begin{pmatrix} \delta & 0 \\ 0 & \delta \end{pmatrix} \middle| \quad \delta \in K^\times\right\}$$

of G is also contained in B. Clearly $Z \cap P = 1$ and $ZP = B$, i.e., B is the semi-direct product of Z and P.

Note that U is contained in P. In fact U is also the commutator subgroup of P. A complement of U in P is the group

$$A = \left\{\begin{pmatrix}\alpha & 0\\ 0 & 1\end{pmatrix} \middle|\ \alpha \in K^\times\right\},$$

which is canonically isomorphic to $K^\times$. Thus P is the semi-direct product of U by A. The action of A on U by conjugation corresponds to the action of $K^\times$ on K^+ by multiplication

$$\begin{pmatrix}\alpha & 0\\ 0 & 1\end{pmatrix}\begin{pmatrix}1 & \beta\\ 0 & 1\end{pmatrix}\begin{pmatrix}\alpha^{-1} & 0\\ 0 & 1\end{pmatrix} = \begin{pmatrix}1 & \alpha\beta\\ 0 & 1\end{pmatrix}.$$

Our method of constructing the representations of G consists of three stages: First of all we use general principles and easily determine the representations of P. Then we make a jump to B and induce characters from B to G. The last and the most difficult stage is to explore those representations of G that do not appear in the former stage. In doing this we shall use the <u>Bruhat's decomposition</u> of G, namely $G = B \cup BwU$. Indeed, if $\gamma \neq 0$, then

$$\begin{pmatrix}\alpha & \beta\\ \gamma & \delta\end{pmatrix} = \begin{pmatrix}\beta-\alpha\gamma^{-1}\delta & \alpha\\ 0 & \gamma\end{pmatrix}\begin{pmatrix}0 & 1\\ 1 & 0\end{pmatrix}\begin{pmatrix}1 & \gamma^{-1}\delta\\ 0 & 1\end{pmatrix}.$$

5. The conjugacy classes of GL(2,K).

Before we start to investigate the irreducible representations of G, we would like to compute their number. It is equal to the number of the conjugacy classes of G. The computation of this number will be done by explicitly giving a representative for each of the conjugacy classes. This will also help us later to give the <u>characters table</u> of G, i.e., the values of the irreducible higher dimensional characters at the conjugacy classes.

An element g of G has two eigenvalues. If one of them belongs to K, then so does the other, since they both satisfy the same quadratic equation, $\det(g-XI) = 0$ over K. All the elements in the conjugacy class of g have the same eigenvalues. There are therefore two possibilities:

(a) The eigenvalues of g belong to K.

In this case g is conjugate over K to a unique matrix in a canonical Jordan form. If both eigenvalues are equal to the same element α of K, then the Jordan form is

$$(1) \qquad c_1(\alpha) = \begin{pmatrix} \alpha & 0 \\ 0 & \alpha \end{pmatrix} \quad \text{or} \quad c_2(\alpha) = \begin{pmatrix} \alpha & 1 \\ 0 & \alpha \end{pmatrix},$$

depending on whether the minimal polynomial of g is different from the characteristic polynomial or equal to it. If the eigenvalues are α,β and $\alpha \neq \beta$, then the Jordan form is

$$(2) \qquad c_3(\alpha,\beta) = \begin{pmatrix} \alpha & 0 \\ 0 & \beta \end{pmatrix}.$$

There are $q-1$ matrices of the form $c_1(\alpha)$, $q-1$ of the form $c_2(\alpha)$ and $\frac{1}{2}(q-1)(q-2)$ of the form $c_3(\alpha,\beta)$.

(b) The eigenvalues of g do not belong to K.

In this case they belong to the unique quadratic extension L of K. Denote by $p(X)$ the characteristic polynomial of g. Then $p(X)$ is irreducible over K and its roots $\alpha,\bar{\alpha}$, which are the eigenvalues of g, are conjugate over K. They are distinct, since K as a finite field is perfect. If we denote $\mathrm{Tr}(\alpha) = \alpha+\bar{\alpha}$ and $N\alpha = \alpha\bar{\alpha}$, then $p(X) = X^2-\mathrm{Tr}(\alpha)K+N\alpha$.

Let v be a non-zero vector in K^2. Then v, gv form a basis for K^2 over K, since otherwise there would exist a $\lambda \in K$ such that $gv = \lambda v$. This λ would then be an eigenvalue of g in K, contrary to our hypothesis. Recalling that $p(g) = 0$ (by the Cayley-Hamilton theorem), we have that the matrix of g, when considered as a linear operator on K^2 with respect to the basis v, gv, is

$$c_4(\alpha) = \begin{pmatrix} 0 & -N\alpha \\ 1 & \mathrm{Tr}\alpha \end{pmatrix}.$$

Thus g is conjugate in G to $c_4(\alpha)$.

Conversely, given an $\alpha \in L-K$, then $c_4(\alpha)$ is a matrix in G with the eigenvalues $\alpha, \bar{\alpha}$. If β is an additional element of $L-K$, then $c_4(\alpha)$ is conjugate to $c_4(\beta)$ if and only if $\beta = \alpha$ or $\beta = \bar{\alpha}$, since then $p(\beta) = 0$.

There are q^2-q elements in $L-K$. Hence there are $\frac{1}{2}(q^2-q)$ matrices of the form $c_4(\alpha)$.

We sum up our results in the following:

Proposition 5.1: The conjugacy classes of G are classified in four families:

(1) $q-1$ classes, represented by $c_1(\alpha)$, with equal eigenvalues in K such that the characteristic polynomial is different from the minimal polynomial;

(2) $q-1$ classes represented by $c_2(\alpha)$, with equal eigenvalues in K such that the characteristic polynomial is equal to the minimal polynomial;

(3) $\frac{1}{2}(q-1)(q-2)$ classes, represented by $c_3(\alpha,\beta)$, with distinct eigenvalues in K;

(4) $\frac{1}{2}(q^2-q)$ classes, represented by $c_4(\alpha)$, with eigenvalues in $L-K$.

Chapter 2. The representations of GL(2,K)

This chapter starts with the representations of P, then investigates the behavior of representations of G that are induced from characters of B, and finally describes the cuspidal representations of G, i.e., those representations that do not appear as components of the induced ones. The chapter ends with Weil's reciprocity law.

6. The representations of P.

We use the method of "small groups" of Wigner in order to determine the representations of P (cf. [2, p. 78]).

First we fix for the rest of these notes a non-unit character ψ of K^+. We consider it also as a character of U. For every $a \in A$ we define a character ψ_a of U by

(1) $$\psi_a(u) = \psi(aua^{-1}), \quad \text{for } u \in U .$$

If $a \neq a'$, then $\psi_a \neq \psi_{a'}$. Indeed, if

$$a = \begin{pmatrix} \alpha & 0 \\ 0 & 1 \end{pmatrix}, \quad a' = \begin{pmatrix} \alpha' & 0 \\ 0 & 1 \end{pmatrix} \quad \text{and} \quad u = \begin{pmatrix} 1 & \beta \\ 0 & 1 \end{pmatrix},$$

then $\psi_a(u) = \psi(\alpha\beta)$, and $\psi_a = \psi_{a'}$ implies that $\psi((\alpha-\alpha')\beta) = 0$ for all $\beta \in K$; hence $\alpha = \alpha'$; and hence $c = c'$. We thus get q-1 distinct representations of U. These, together with the unit representation of U, are all the characters of U since $|U| = q$.

Every character χ of A can be lifted to a character $\tilde{\chi}$ of P defined by $\tilde{\chi}(ua) = \chi(a)$. The q-1 distinct characters $\tilde{\chi}$ of P obtained in this way are all the characters of P since $(P:P') = (P:U) = q-1$.

In order to find the higher dimensional representations of P, we induce ψ from U to P and claim

$$(2) \qquad \mathrm{Res}_U^P \mathrm{Ind}_U^P \psi = \bigoplus_{a \in A} \psi_a \quad .$$

Indeed, for every $a \in A$ we define a function $f_a \in \mathrm{Ind}_U^P V_\psi$ by

$$f_a(a') = \begin{cases} 1 & \text{if } a = a' \\ 0 & \text{if } a \neq a' \end{cases} \quad , \text{ where } a' \in A.$$

Then f_a is an eigenvector of U that belongs to the eigenvalue ψ_a. In order to prove this claim, we have to show that $f_a(pu) = \psi_a(u)f_a(p)$ for every $p \in P$ and every $u \in U$. Writing $p = u'a'$ with u' in U and a' in A and using the identity $f_a(u'p') = \psi(u')f_a(p')$ for $p' \in P$, we see that it suffices to show that

$$(3) \qquad f_a(a'u) = \psi_a(u)f_a(a') \; .$$

Indeed

$$f_a(a'u) = f_a(a'ua'^{-1}a') = \psi(a'ua'^{-1})f_a(a') = \psi_{a'}(u)f_a(a') \; .$$

The right-hand side is equal to zero if $a \neq a'$ and equal to $\psi_a(u)f_a(a')$ if $a = a'$; hence (3) is true in both cases. Thus the vector f_a generates the one-dimensional space V_{ψ_a}.

If we let a vary, we get $q-1$ linearly independent vectors f_a of the $(q-1)$-dimensional vector space $\mathrm{Ind}_U^P V_\psi$. Hence $\mathrm{Res}_U^P \mathrm{Ind}_U^P V_\psi = \bigoplus_{a \in A} V_{\psi_a}$ as U-modules. This proves (2).

As a consequence of (2) we prove the following fundamental:

<u>Theorem 6.1</u>: The group P has q irreducible representations:

(a) $(q-1)$ of them are one-dimensional; they are the lifting of the characters of A;

(b) one $(q-1)$-dimensional representation which is $\pi = \mathrm{Ind}_U^P \psi$.

Proof: We only have to prove (b). First note that $\dim \mathrm{Ind}_U^P\psi = (P:U) = q-1$. Second, by the Frobenius reciprocity theorem and by (2)

$$(\mathrm{Ind}_U^P\psi,\ \mathrm{Ind}_U^P\psi)_P = (\psi, \bigoplus_{a\in A}\psi_a) = 1 ;$$

hence $\mathrm{Ind}_U^P\psi$ is an irreducible character of P.

In order to prove that there is no additional representation of P, one can observe that

$$\sum_{a\in A} (\dim\psi_a)^2 + (\dim \mathrm{Ind}_U^P\psi)^2 = (q-1)+(q-1)^2 = |P| .$$

We would however also like to prove the last assertion without using the finiteness of G. In order to do this, note first that one can in fact replace ψ in (2) by $\psi_{a'}$ and have

$$\text{(4)} \qquad \mathrm{Res}_U^P\ \mathrm{Ind}_U^P\psi_{a'} = \bigoplus_{a\in A}\psi_a .$$

Hence we can prove, as before, that $\mathrm{Ind}_U^P\psi_{a'}$ is an irreducible representation of P. Further, by (4)

$$(\mathrm{Ind}_U^P\psi,\ \mathrm{Ind}_U^P\psi_{a'}) = (\psi, \bigoplus_{a\in A}\psi_a) = 1 .$$

Hence

$$\text{(5)} \qquad \mathrm{Ind}_U^P\psi = \mathrm{Ind}_U^P\psi_{a'} .$$

Now let σ be an arbitrary irreducible representation of P and consider $\mathrm{Res}_U^P\sigma$. If there exists an $a' \in A$ such that $\psi_{a'}$ appears in $\mathrm{Res}_U^R\sigma$, then by (5)

$$(\sigma,\ \mathrm{Ind}_U^P\psi) = (\mathrm{Res}_U^P\sigma,\ \psi_{a'}) > 0 .$$

Hence $\sigma = \mathrm{Ind}_U^P\psi$, since both representations are irreducible. Otherwise, $\mathrm{Res}_U^P\sigma$ is a multiple of the unit character of U, i.e., $\sigma(u)v = v$ for every $v \in V_\sigma$. Consider therefore $\mathrm{Res}_A^P V_\sigma$. It decomposes into linear A-subspaces since A is abelian. In particular there exists a vector $0 \neq v \in V_\sigma$ and a character of A, say χ, such that $\sigma(a)v = \chi(a)v$ for every $a \in A$. Hence, if $u \in U$, then $\sigma(au)v = \chi(a)v$. It follows that $\sigma = \tilde{\chi}$. //

Remarks: (1) Note that we have also proved that our description of the representation of P is independent of the choice of ψ. (2) The distinguished representation $\pi = \mathrm{Ind}_U^P\psi$ will play an important role in the sequel. This is the reason for reserving the letter π for it.

7. The representations of B.

We have already mentioned that the commutator of B is U (see section 4). Moreover, B is the semi-direct product of D and U. The group D is canonically isomorphic to $K^\times \times K^\times$. Hence every pair (μ_1,μ_2) of characters of $K^\times$ defines a unique character μ of B which is given by the formula

$$(1) \qquad \mu\begin{pmatrix} \alpha & * \\ 0 & \delta \end{pmatrix} = \mu_1(\alpha)\mu_2(\delta) \qquad \alpha,\delta \in K^\times .$$

Conversely, to every character μ of B there corresponds a pair of characters (μ_1,μ_2) of $K^\times$ such that (1) holds. Thus the $(q-1)^2$ characters of B given by (1) are all the characters of B.

An easy computation shows that the normalizer of D in G is generated by D and w. Indeed we have

$$w\begin{pmatrix} \alpha & 0 \\ 0 & \delta \end{pmatrix}w^{-1} = \begin{pmatrix} \delta & 0 \\ 0 & \alpha \end{pmatrix} .$$

For every character μ of B given by (1), we define a character μ_w of B by

$$\mu_\omega\begin{pmatrix} \alpha & * \\ 0 & \delta \end{pmatrix} = \mu_1(\delta)\mu_2(\alpha) .$$

Then $\mu_w(d) = \mu(wdw^{-1})$ for every $d \in D$ and

$$\mu_w = \mu \iff \mu_1 = \mu_2 .$$

In order to find the higher dimensional representations of B, recall that B is also the semi-direct product of Z by P. The abelian group Z has $q-1$ characters χ. Each of them can be extended to a character $\tilde{\chi}$ of B by

$$\tilde{\chi}(zp) = \chi(z) \qquad z \in Z, \quad p \in P .$$

Also, composing the canonical map $B \to P$ with kernel Z, with the representation π of B (see Theorem 6.1), we get an irreducible representation $\tilde{\pi}$ of B of dimension $q-1$:

$$\tilde{\pi}(zp) = \pi(p) \qquad z \in Z, \quad p \in P .$$

The tensor product

$$(2) \qquad (\tilde{\chi}\otimes\tilde{\pi})(zp) = \chi(z)\pi(p) \qquad z \in Z, \quad p \in P$$

is an irreducible $(q-1)$-dimensional representation of B whose restriction to Z is χ. Varying χ on all the characters of Z, we get $q-1$ $(q-1)$-dimensional representations of B. These, together with the $(q-1)^2$ characters of B, are all the irreducible representations of B since

$$(q-1)^2 + (q-1)(q-1)^2 = q(q-1)^2 = |B| .$$

We have therefore proved the following:

<u>Theorem 7.1</u>: The group B has

(a) $(q-1)^2$ characters given by (1); and

(b) $q-1$ $(q-1)$-dimensional representations given by (2).

8. Inducing characters from B to G.

As a first step toward the determination of the irreducible representations of G, we investigate those that appear as components of $\mathrm{Ind}_B^G\mu$ where μ is a character of B. In order to shorten the notation, we make the convention

$$\hat{\mu} = \mathrm{Ind}_B^G \mu$$

and stick to it for the rest of these notes. The dimension of $\hat{\mu}$ is $q+1$. Our task in this section is to determine the connection between μ and $\hat{\mu}$.

In order to do this we define the <u>Jacquet Module</u> of a representation ρ of G as

$$J(V_\rho) = \{v \in V_\rho \mid \rho(u)v = v \quad \text{for every } u \in U\} .$$

The fact that U is normal in B implies that B acts on $J(V_\rho)$. Indeed, if $v \in J(V_\rho)$, $b \in B$ and $u \in U$, then $b^{-1}ub \in U$; hence

$$\rho(u)\rho(b)v = \rho(b)\rho(b^{-1}ub)v = \rho(b)v .$$

It might happen however that $J(V_\rho)$ is not a G-space.

If ρ_1,ρ_2 are two representations of ρ, then clearly

$$J(V_{\rho_1} \oplus V_{\rho_2}) = J(V_{\rho_1}) \oplus J(V_{\rho_2}) .$$

In particular we have:

<u>Lemma 8.1</u>: If μ is a character of B, then $\dim J(V_{\hat{\mu}}) = 2$.

<u>Proof</u>: By definition $J(V_{\hat{\mu}})$ consists of all the functions $f: G \to \mathbb{C}$ that satisfy

$$f(bg) = \mu(b)f(g) \quad \text{and} \quad f(bu) = f(b)$$

for all $b \in B$, $g \in G$ and $u \in U$. In particular,

$$f(b) = \mu(b)f(1) \quad \text{and} \quad f(bwu) = \mu(b)f(w).$$

Using the Bruhat decomposition $G = B \cup BwU$, this implies that f is determined by its values in 1 and in w, where it can be arbitrary. It follows that $\dim J(V_{\hat{\mu}}) = 2$, and a canonical basis for $J(V_{\hat{\mu}})$ is the two functions f_1, f_2 satisfying

$$(1)\qquad \begin{array}{ll} f_1(1) = 1 & f_1(w) = 0 \\ f_2(1) = 0 & f_2(w) = 1 \end{array} .$$

A supplement to Lemma 8.1 is:

Lemma 8.2: If μ is a character of B, then B operating on $J(V_{\hat{\mu}})$ has two eigenvectors f_1, f_2 (defined by (1)) that correspond to the eigenvalues μ and μ_w, respectively.

In detail

$$(2)\qquad \hat{\mu}(b)f_1 = \mu(b)f_1 \quad \text{and} \quad \hat{\mu}(b)f_2 = \mu_w(b)f_2$$

for every $b \in B$.

Proof: We have only to prove that both sides of the equalities (2) coincide in 1 and in w.

Indeed, for f_1 we have $(\hat{\mu}(b)f_1)(1) = f_1(b) = \mu(b)f_1(1)$. Also by the Bruhat decomposition there exists for every $b \in B$ elements $b_1 \in B$ and $u \in U$ such that $wb = b_1wu$. Hence

$$(\hat{\mu}(b)f_1)(w) = f_1(wb) = f_1(b_1wu) = \mu(b_1)f_1(w) = 0 = \mu(b)f_1(w).$$

For f_2 we have

$$(\hat{\mu}(b)f_2)(1) = f_2(b) = \mu(b)f_2(1) = 0 = \mu_w(b)f_2(1) .$$

A difficulty arises in calculating $(\hat{\mu}(b)f_2)(w)$. We overcome this by first considering $d \in D$. Then

$$(3)\qquad (\hat{\mu}(d)f_2)(w) = f_2(wd) = f_2(wdww) = \mu(wdw)f_2(w) = \mu_w(d)f_2(w) .$$

In general we know from Lemma 8.1 that f_1 and f_2 generate $J(V_{\hat{\mu}})$. Hence for every $b \in B$ there exist $\alpha_1(b), \alpha_2(b) \in \mathbb{C}$ such that

$$(4)\qquad \hat{\mu}(b)f_2 = \alpha_1(b)f_1 + \alpha_2(b)f_2 .$$

Calculating both sides of (4) at 1, we obtain $\alpha_1(b) = 0$; hence $\hat{\mu}(b)f_2 = \alpha_2(b)f_2$. It follows that $\alpha_2(b_1b_2) = \alpha_2(b_1)\alpha_2(b_2)$ for $b_1,b_2 \in B$, i.e., α_2 is a character of B. Hence, if

$$b = \begin{pmatrix} \alpha & * \\ 0 & \delta \end{pmatrix} \quad \text{and} \quad d = \begin{pmatrix} \alpha & 0 \\ 0 & \delta \end{pmatrix},$$

then

$$(\hat{\mu}(b)f_2)(w) = \alpha_2(b)f_2(w) = \alpha_2(d)f_2(w) = (\hat{\mu}(d)f_2)(w) =$$

$$= \mu_w(d)f_2(w) = \mu_w(b)f_2(w). \qquad //$$

The importance of the Jacquet modules for our investigation lies in the following:

Lemma 8.3: Let ρ be a representation of G. Then $J(V_\rho) \neq 0$ if and only if there exists a character μ of B, such that $(\rho,\hat{\mu}) \neq 0$.

Proof: Suppose that $J(V_\rho) \neq 0$. Then $J(V_\rho)$ can be considered as a non-trivial B/U-space via ρ. But B/U is abelian; hence $J(V_\rho)$ splits into a direct sum of one-dimensional B/U subspaces. It follows that there exists a character μ of B and a non-zero element $v \in J(V_\rho)$ such that $\rho(b)v = \mu(b)v$ for every $b \in B$. Hence $(\mathrm{Res}_B^G\rho, \mu) \neq 0$. By the Frobenius reciprocity theorem $(\rho,\hat{\mu}) \neq 0$, and half of the lemma is thus proved.

Now suppose that $(\rho,\hat{\mu}) \neq 0$. Then arguing backwards we find that there exists a non-zero element $v \in V_\rho$ such that $\rho(b)v = \mu(b)v$ for every $b \in B$. Now recall that μ is trivial on U since U is the commutator of G. Hence $v \in J(V_\rho)$, which is therefore not zero. //

Corollary 8.4: Let ρ be an irreducible representation of G. Then $J(V_\rho) \neq 0$ if and only if there exists a character μ of B such that $\rho \lesssim \hat{\mu}$.

Corollary 8.5: If μ is a character of B, then $\hat{\mu}$ has at most two irreducible components.

Proof: Let $\hat{\mu} = \rho_1 \oplus \ldots \oplus \rho_r$ be a decomposition of $\hat{\mu}$ into irreducible components. Then $J(V_{\hat{\mu}}) = J(V_{\rho_1}) \oplus \ldots \oplus J(V_{\rho_r})$. By Corollary 8.4 $J(V_{\rho_i}) = 0$ for $i = 1,\ldots,r$. Hence the dimension of the right-hand side is $\geq r$. On the other hand, $\dim J(V_{\hat{\mu}}) = 2$ by Lemma 8.1. Hence $r \leq 2$. //

The next few lemmas give the exact information about the possible two components of $\hat{\mu}$.

Lemma 8.6: If μ is a character of B and $\mu = \mu_w$, then $\hat{\mu}$ has a one-dimensional component.

Proof: The assumption implies that μ corresponds to a pair (μ_1, μ_2) of characters $K^\times$, i.e., if $b = \begin{pmatrix} \alpha & * \\ 0 & \delta \end{pmatrix}$, then $\mu(b) = \mu_1(\alpha)\mu_1(\delta)$. It follows that $\mu(b) = \mu_1(\det b)$. Now define a function $f\colon G \to \mathbb{C}$ by $f(g) = \mu_1(\det g)$. Then $f(bg) = \mu(b)f(g)$, that is, f belongs to $V_{\hat{\mu}}$. Moreover, $(\hat{\mu}(s)f)(g) = f(gs) = \mu_1(\det s)f(g)$ for $s,g \in G$. It follows that f is an eigenvector of G that belongs to the eigenvalue $u_1 \circ \det$. //

Lemma 8.7: If μ is a character of B, then $\hat{\mu}$ has at most one one-dimensional component.

Proof: Assume that $\hat{\mu}$ has two one-dimensional components, χ_1 and χ_2. Then by Corollary 8.5 they are all the components of $\hat{\mu}$, i.e., $\hat{\mu} = \chi_1 \oplus \chi_2$. It follows that $q+1 = \dim \hat{\mu} = 2$, which is a contradiction. //

Lemma 8.8: If μ is a character of B, then

$$\mathrm{Res}_P V_{\hat{\mu}} \equiv \mathrm{Res}_P J(V_{\hat{\mu}}) \oplus V_\pi \quad .$$

Proof: $J(V_{\hat{\mu}})$ is a two-dimensional B-subspace of $V_{\hat{\mu}}$; in particular $J(V_{\hat{\mu}})$ is a P-subspace of $V_{\hat{\mu}}$. Let V be a P-complement to $J(V_{\hat{\mu}})$ in $V_{\hat{\mu}}$. Then $\dim V = p-1$. Further V has no one-dimensional P-subspace; indeed otherwise, there would exist a non-zero element $v \in V$ and a character χ of P such that $\hat{\mu}(p)v = \chi(p)v$ for every $p \in P$. In particular we would have for $u \in U$ that $\hat{\mu}(u)v = v$ since $U = P'$. It follows that $v \in J(V_{\hat{\mu}})$; hence $v = 0$, which is a contradiction. Therefore, by Theorem 6.1 V cannot have irreducible P-subspaces of dimension $< p-1$. Hence V is irreducible and isomorphic to the unique irreducible P-space V_π of dimension $p-1$. //

<u>Lemma 8.9</u>: If μ is a character of B and $\hat{\mu}$ is reducible, then:

(a) $\hat{\mu}$ has a one-dimensional component;

(b) $\mu = \mu_w$.

<u>Proof</u>:

(a) Let $V_{\hat{\mu}} = V \oplus V'$ be a non-trivial G-decomposition of V. By Lemma 8.8 we can assume, without loss of generalization, that $V_\pi \cap V \neq 0$. Then $V_\pi \subseteq V$ since V_π is an irreducible P-space. On the other hand, by Lemma 8.3 $0 \neq J(V) \subseteq J(V_{\hat{\mu}}) \cap V$. Hence $V_\pi \neq V$. It follows that $\dim V = q$ and hence $\dim V' = 1$.

(b) We have proved that there exists a character χ of G and a non-zero function $f\colon G \to \mathbb{C}$ such that

(5) $$f(bg) = \mu(b)f(g) \text{ and } f(gs) = \chi(s)f(g) \text{ for every } b \in B \text{ and } g,s \in G .$$

<u>Claim</u>: $f(1) \neq 0$. Indeed, assume that $f(1) = 0$ and let $g \in G$. Then there exists a positive integer n such that $g^n = 1$. Hence $0 = f(1) = f(g\cdot g^{m-1}) = \chi(g^{m-1})f(g)$. It follows that $f(g) = 0$ since $\chi(g^{m-1}) \neq 0$. This is a contradiction.

Let $d \in D$; then $\mu(d)f(1) = f(d) = \chi(d)f(1)$. Hence $\mu(d) = \chi(d)$. It follows that $\mu_w(d) = \mu(wdw) = \chi(wdw) = \chi(w^2)\chi(d) = \mu(d)$. Hence $\mu_w = \mu$. //

<u>Lemma 8.10</u>: Let μ and μ' be two distinct characters of B. Then $(\mu,\hat{\mu}') \neq 0$ if and only if $\mu' = \mu_w$.

<u>Proof</u>: Suppose first that $(\hat{\mu},\hat{\mu}') \neq 0$. Then there exists an irreducible representation ρ of G such that $\rho \leqslant \hat{\mu}$ and $\rho \leqslant \hat{\mu}'$. It follows by Corollary 8.4 that $J(V_\rho) \neq 0$. By the proof of Lemma 8.3 we know that B operating on $J(V_\rho)$ has an eigenwert χ. In addition, by Lemma 8.2 we know that B

operating on $J(V_{\hat{\mu}})$ has exactly two eigenvalues μ and μ_w. It follows that $\chi = \mu$ or $\chi = \mu_w$ since $J(V_\rho) \subseteq J(V_{\hat{\mu}})$. Similarly $\chi = \mu'$ or $\chi = \mu'_w$, but $\mu \neq \mu'$; hence $\mu' = \mu_w$.

Conversely, suppose that $\mu' = \mu_w$. Then arguing backwards we have that $(\mu', \mathrm{Res}_B\hat{\mu}) \neq 0$. Hence by the Frobenius reciprocity theorem $(\hat{\mu}', \hat{\mu}) \neq 0$. //

<u>Lemma 8.11</u>: Let μ and μ' be two distinct characters of B. Then $\hat{\mu} = \hat{\mu}'$ if and only if $\mu' = \mu_w$.

<u>Proof</u>: Suppose that $\mu' = \mu_w$. Then $\mu \neq \mu_w$ and $\mu' \neq \mu'_w$. It follows that $\hat{\mu}$ and $\hat{\mu}'$ are irreducible by Lemma 8.8. Also $(\hat{\mu}, \hat{\mu}') \neq 0$ by Lemma 8.10. Hence $\hat{\mu} = \hat{\mu}'$.

The other direction of the lemma follows directly from Lemma 8.10. //

Summing up the lemmas in this section we obtain:

<u>Theorem 8.12</u>: Let μ and μ' be characters of B and let $\hat{\mu} = \mathrm{Ind}_B^G \mu$; then:

(a) $\dim \hat{\mu} = q+1$;

(b) $\hat{\mu}$ has at most two irreducible components;

(c) $\hat{\mu}$ is irreducible if and only if $\mu \neq \mu_w$;

(d) if $\hat{\mu}$ is reducible, it decomposes into a direct sum of a one-dimensional and a q-dimensional representation;

(e) $\hat{\mu}' = \mu$ if and only if $\mu' = \mu$ or $\mu' = \mu_w$.

Now consider a character μ of B that corresponds to the pair of characters (μ_1, μ_2) of $K^\times$. There are two possibilities:

(a) $\mu = \mu_w$, i.e., $\mu_1 = \mu_2$. In this case $\hat{\mu}$ is the direct sum of a one-dimensional representation $\rho'_{(\mu_1,\mu_2)}$ and an irreducible q-dimensional representation $\rho_{(\mu_1,\mu_2)}$. $K^\times$ has q-1 characters; hence in this way we obtain q-1 characters of G and q-1 q-dimensional irreducible representations of G.

(b) $\mu \neq \mu_w$, i.e., $\mu_1 \neq \mu_2$. In this case $\hat{\mu}$ is an irreducible representation of dimension q+1 and we denote it by $\rho_{(\mu_1,\mu_2)}$. The number of these μ is equal to the number of characters of B, i.e., $(q-1)^2$ minus the characters of type (a), i.e., $q-1$. Further μ and μ_w induce the same representation. Hence we obtain in this way $\frac{1}{2}(q-1)(q-2)$ irreducible representations of G of dimension $q+1$.

We have therefore proved:

<u>Theorem 8.13</u>: The irreducible representations of G, which are components of induced representations of the form $\mathrm{Ind}_B^G\mu$ where μ is a character of B, split up into the following classes:

(1) $q-1$ one-dimensional representations, $\rho'_{(\mu_1,\mu_1)}$;

(2) $q-1$ q-dimensional representations, $\rho_{(\mu_1,\mu_1)}$;

(3) $\frac{1}{2}(q-1)(q-2)$ $(q+1)$-dimensional representations, $\rho_{(\mu_1,\mu_2)}$.

If χ is a character of G, then χ is a component of $\mathrm{Ind}_B^G\mathrm{Res}_B^G\chi$. It follows by Theorem 8.13 that G has exactly $q-1$ characters. Hence $(G:G') = q-1$. The subgroup $SL(2,K) = \{g\in G \mid \det g = 1\}$ is normal and $G/SL(2,K) \cong K^\times$. Hence,

<u>Corollary 8.14</u>: $SL(2,K)$ is the commutator subgroup of $GL(2,K)$.

9. The Schur algebra of $\mathrm{Ind}_B^G\mu$.

Let μ be a character of B and let $\hat{\mu} = \mathrm{Ind}_B^G\mu$. Bruhat's decomposition of G implies that $|B\backslash G/B| = 2$. Hence, by Corollary 3.2 $(\hat{\mu},\hat{\mu}) \leq 2$. Since $(\hat{\mu},\hat{\mu}) \geq 1$, there are only two possibilities: either $(\hat{\mu},\hat{\mu}) = 1$, in which case $\hat{\mu}$ is irreducible; or $(\hat{\mu},\hat{\mu}) = 2$. We also know that if $\hat{\mu} = \bigoplus_{i=1}^{r} n_i\rho_i$ is the canonical decomposition of $\hat{\mu}$, then $(\hat{\mu},\hat{\mu}) = \sum_{i=1}^{r} n_i^2$. Since 2 can be decomposed into a sum of squares only in the form $2 = 1^2+1^2$, it follows that $(\hat{\mu},\hat{\mu}) = 2$ implies that $\hat{\mu}$ decomposes into a direct sum of two non-isomorphic representations. We have therefore proved:

Theorem 9.1: Let μ be a character of B and let $\hat{\mu} = \mathrm{Ind}_B^G \mu$. Then either $(\hat{\mu},\hat{\mu}) = 1$, in which case $\hat{\mu}$ is irreducible, or $(\hat{\mu},\hat{\mu}) = 2$, in which case $\hat{\mu}$ decomposes into a direct sum of two non-isomorphic representations.

Obviously, Theorem 9.1 is also a consequence of Theorem 8.11. However, the proof given above is independent of Theorem 8.11.

10. The dimension of cuspidal representations.

Irreducible representations of G that are not components of $\hat{\mu}$, with μ a character of B, are said to be cuspidal. By Corollary 8.4 an irreducible representation ρ of G is cuspidal if and only if $J(V_\rho) = 0$. Comparing Proposition 5.1 with Theorem 8.12, we find that G has $\frac{1}{2}(q^2-q)$ cuspidal representations, exactly as the number of conjugacy classes of the form $c_4(\alpha)$.

We delay a further explanation of this phenomenon to section 15 and concentrate in this section on proving that all cuspidal representations have dimension $q-1$.

The first step toward this goal is:

Lemma 10.1: Let ρ be a cuspidal representation of G. Then $\mathrm{Res}_P \rho = r\pi$ for some positive integer r. In particular $\dim \rho = r(q-1)$ is a multiple of $q-1$.

Proof: The P-space $\mathrm{Res}_P^G V_\rho$ cannot have one-dimensional components. Indeed, otherwise there would exist a non-zero vector $v \in V_\rho$ and a character χ of P such that $\rho(p)v = \chi(p)v$ for every $p \in P$. In particular we would have that $\rho(u)v = v$ for every $u \in U$, i.e., $v \in J(V_\rho)$; thus $J(V_\rho) \neq 0$, contrary to the assumption that ρ is cuspidal (cf. the proof of Lemma 8.8). By Theorem 6.1 $\mathrm{Res}_P^G \rho$ must be a multiple of π. //

Proposition 10.2:

(a) Let ρ be a cuspidal representation. Then $\mathrm{Res}_P \rho = \pi$ and $\dim \rho = q+1$.

(b) Conversely, if ρ is a representation of G such that $\mathrm{Res}_P \rho = \pi$, then ρ is cuspidal.

Proof:

(a) Using the formula $|G| = \Sigma(\dim\sigma)^2$ where σ runs over the irreducible representations of G, and by Theorem 8.13 we have

$$(q-1)^2 q(q+1) \geq (q-1)\cdot 1^2+(q-1)q^2+\tfrac{1}{2}(q-1)(q-2)(q+1)^2+\textstyle\sum'(\dim\sigma)^2 ,$$

where the prime indicates that σ runs over the cuspidal representations. By Lemma 10.1 there exists for every σ a positive integer $r(\sigma)$ such that $\dim\sigma = (q-1)r(\sigma)$. Hence,

$$\tfrac{1}{2}(q^2-q) \geq \textstyle\sum' r(\sigma) .$$

The number of the summands on the right-hand side is equal to $\frac{1}{2}(q^2-q)$. Hence $r(\sigma) = 1$ and (a) follows from Lemma 10.1.

(b) The representation π is irreducible; hence ρ is irreducible too. Also, if μ is a character of B, then by Theorem 8.13 the components of $\hat{\mu}$ have the dimensions 1, q or $q+1$. However, $\dim\rho = \dim\pi = q-1$; hence ρ is not equal to any of them (i.e., ρ is cuspidal). //

The proof of Proposition 10.2(a) relies heavily on the fact that K is a finite field. We now give another proof that will be independent of this fact.

Let $\hat{\pi} = \mathrm{Ind}_B^G \pi = \mathrm{Ind}_U^G \psi$. Then by Proposition 3.1 $\mathrm{End}_{\mathbb{C}[G]} V_{\hat{\pi}}$ is isomorphic to the algebra A of all functions $F:G \to \mathbb{C}$ satisfying

$$F(u_1 g u_2) = \psi(u_1 u_2)F(g) \quad \text{for} \quad u_1,u_2 \in U \text{ and } g \in G ,$$

and where multiplication between two functions $F_1,F_2 \in A$ is given by the formula

$$(F_1 * F_2)(g) = \frac{1}{(G:U)} \sum_{s \in G} F_1(gs^{-1})F_2(s)$$

(see (8) of section 3). We shall show that A is abelian. This implies that $\hat{\pi}$ has no multiple components (see section 1). If ρ is a cuspidal representation, then by Lemma 10.1 there exists a positive integer r such that $(\mathrm{Res}_P \rho, \pi) = r$ and $\dim \rho = r(q-1)$. Hence by the Frobenius reciprocity theorem $r = (\rho, \hat{\pi}) = 1$, and our contention is proved.

Our method of proving that A is abelian is indirect. We shall define an involution on A, i.e., a map $F \to F'$ such that $(F_1 * F_2)' = F_2' * F_1'$. We further prove that $F = F'$ for every $F \in A$. Hence $F_1 * F_2 = F_2 * F_1$.

We start by defining an involution $g \to g'$ on G: If $g = \begin{pmatrix} \alpha & \beta \\ \gamma & \delta \end{pmatrix}$, we let $g' = \begin{pmatrix} \delta & \beta \\ \gamma & \alpha \end{pmatrix}$. Then $(g_1 g_2)' = g_2' g_1'$, $g'' = g$ and $g = g'$ if g is symmetric with respect to the second diagonal. In particular $u' = u$ for every $u \in U$. We continue by defining for an element $F \in A$ a function $F': G \to \mathbb{C}$ by $F'(g) = F(g')$. Then F belongs to A. In order to prove that $F = F'$, it suffices to show that F and F' coincide on representatives of the double classes $U \backslash G / U$. Indeed, by Bruhat's decomposition $G = B \cup BwU$ and since $B = UD$, we have that the above representatives are either of the form

$$\text{(a)} \quad \begin{pmatrix} 0 & \beta \\ \gamma & 0 \end{pmatrix}, \qquad \text{or of the form (b)} \quad \begin{pmatrix} \alpha & 0 \\ 0 & \delta \end{pmatrix}.$$

Clearly F and F' coincide on the matrices (a) and on the matrices (b) in case $\alpha = \delta$. In order to prove that F and F' coincide also on the matrices (b) in case $\alpha \neq \delta$, it suffices to show that F (and hence also F') vanishes on them. Indeed, acting with F on both sides of the identity

$$\begin{pmatrix} 1 & \beta \\ 0 & 1 \end{pmatrix}\begin{pmatrix} \alpha & 0 \\ 0 & \delta \end{pmatrix} = \begin{pmatrix} \alpha & 0 \\ 0 & \delta \end{pmatrix}\begin{pmatrix} 1 & \alpha^{-1}\delta\beta \\ 0 & 1 \end{pmatrix},$$

we have

$$\psi(\beta) F\begin{pmatrix} \alpha & 0 \\ 0 & \delta \end{pmatrix} = F\begin{pmatrix} \alpha & 0 \\ 0 & \delta \end{pmatrix} \psi(\alpha^{-1}\delta\beta) \quad .$$

If $F\begin{pmatrix} \alpha & 0 \\ 0 & \delta \end{pmatrix} \neq 0$, then $\psi(\beta) = \psi(\alpha^{-1}\delta\beta)$; hence $\psi(\beta(1-\alpha^{-1}\delta)) = 1$ for every $\beta \in K$. It follows that ψ is the unit character of K^+, which is a contradiction.

This completes the alternative proof of Proposition 10.2(a). Note that we have actually also proved the following:

Proposition 10.3: $\mathrm{Ind}_U^G\psi$ has no multiple components.

11. The description of GL(2,K) by generators and relations.

We need this description for an explicit presentation of the cuspidal representations.

Let

$$w' = \begin{pmatrix} 0 & 1 \\ -1 & 0 \end{pmatrix}, \quad z = \begin{pmatrix} 1 & 1 \\ 0 & 1 \end{pmatrix} \quad \text{and} \quad s = w'z .$$

Then we have the following relations between w' and the elements of B:

(1) $$w'\begin{pmatrix} \alpha & 0 \\ 0 & \beta \end{pmatrix}w'^{-1} = \begin{pmatrix} \beta & 0 \\ 0 & \alpha \end{pmatrix},$$

(2) $$w'^2 = \begin{pmatrix} -1 & 0 \\ 0 & -1 \end{pmatrix},$$

(3) $$s^3 = 1 .$$

Proposition 11.1: GL(2,K) is the free group generated by B and w' with (1), (2) and (3) as the defining relations.

Proof: Denote by G the free group generated by B, w' with the above defining relations. Then there exists a unique epimorphism θ of G onto GL(2,K) which is the identity on B and maps w' onto itself. We have to prove that its kernel consists of 1.

Claim: For every $b \in B-D$ there exist $b_1, b_2 \in B$ such that $w'bw' = b_1w'b_2$.

Indeed, if $b = \begin{pmatrix} \alpha & \beta \\ 0 & \delta \end{pmatrix}$ and $\beta \neq 0$, then

$$b = \begin{pmatrix} 1 & 0 \\ 0 & \delta\beta^{-1} \end{pmatrix}\begin{pmatrix} 1 & 1 \\ 0 & 1 \end{pmatrix}\begin{pmatrix} \alpha & 0 \\ 0 & \beta \end{pmatrix} = d'zd'' \quad ;$$

also $w'zw'zw'z = 1$ by (3); hence

$$w'zw' = z^{-1}w'^{-1}z^{-1} = z^{-1}\begin{pmatrix} -1 & 0 \\ 0 & -1 \end{pmatrix}w'z^{-1}$$

by (2). It follows that

$$w'bw' = (w'd'w'^{-1})w'zw'(w'^{-1}d''w') =$$

$$(w'd'w'^{-1})z^{-1}\begin{pmatrix} -1 & 0 \\ 0 & -1 \end{pmatrix}w'z^{-1}(w'^{-1}d''w') = b_1w'b_2 \quad ,$$

by (1).

Next note that if $d \in D$, then $w'dw' = (w'dw'^{-1})w'^2 \in B$ by (1) and (2).

Now let $g \neq 1$ be in the kernel of θ. Then $g \notin B$; hence g can be written, for example, as $g = w'b_1w'b_2\ldots w'b_r$, where $b_i \in B$. If $r \geq 2$, then either $w'b_2w' = b_1' \in B$, if $b_1 \in D$, or $w'b_1w' = b_1'w'b_2'$ with $b_1', b_2' \in D$ if $b_1 \in B\text{-}D$. In any case one can rewrite g as $g = b_2''w'\ldots w'b_r$. By a repeated application of this procedure one finally proves that $g \in Bw'B$, i.e.,

$$g = \begin{pmatrix} \alpha' & \beta' \\ 0 & \delta \end{pmatrix}w'\begin{pmatrix} \alpha & \beta \\ 0 & \delta \end{pmatrix} .$$

The right-hand side is mapped by θ to an element $\begin{pmatrix} * & * \\ -\delta'\alpha & 0 \end{pmatrix}$ of $GL(2,K)$. But g is mapped to 1. Hence $\delta'\alpha = 0$, which is a contradiction. //

12. Non-decomposable characters of $L^\times$.

We have denoted by L the unique quadratic extension of K. It has q^2 elements. If α is an element of L, then $\bar{\alpha}$ denotes its unique conjugate over K. Then the function $N\alpha = \alpha\bar{\alpha}$ is the norm map from L to K. It is multiplicative.

<u>Lemma 12.1</u>: N is surjective.

Proof: The Galois group of L over K is generated by the Frobenius automorphism $\alpha \to \alpha^q$. Hence $N\alpha = \alpha^{q+1}$. The restriction of N to $L^\times$ is a homomorphism into $K^\times$. It follows that the kernel E of this homomorphism consists of $q+1$ elements. Hence, its image consists of $\frac{q^2-1}{q+1} = q-1$ elements, exactly as many elements as $K^\times$ has. //

We note that Lemma 12.1 also follows from the fact that the Brauer group of K is trivial. This proof of the lemma is not so elementary as the one we gave, but it is independent of counting elements.

Corollary 12.2 (Hilbert's Satz 90): If β is an element of $L^\times$ such that $N_\beta = 1$, then there exists an $\alpha \in L$ such that $\alpha\overline{\alpha}^{-1} = \beta$.
Proof: The map $h\colon L^\times \to E$ defined by $h(\delta) = \alpha\overline{\alpha}^{-1}$ is a homomorphism. Its kernel is $K^\times$. Hence the image of h has $(q^2-1)(q-1)^{-1}$ elements, exactly as many as E has. //

Let χ be a character of $K^\times$. Composing χ with the norm map N from $L^\times$ into $K^\times$, we obtain a character $\tilde{\chi}$ of $L^\times$.

$$(1) \qquad \tilde{\chi}(\alpha) = \chi(N\alpha) \qquad \alpha \in L^\times .$$

$\tilde{\chi}$ is said to be decomposable. If ν is an arbitrary character of $L^\times$, then $\overline{\nu}$ denotes its conjugate over K, i.e., $\overline{\nu}(\alpha) = \nu(\overline{\alpha})$ for $\alpha \in L^\times$.

Lemma 12.3: A character ν of $L^\times$ is decomposable if and only if $\nu = \overline{\nu}$.
Proof: If there exists a character χ of $\tilde{K}$ such that $\nu = \tilde{\chi}$, then certainly $\nu(\alpha) = \nu(\overline{\alpha})$ for every $\alpha \in K^\times$. Conversely, if $\nu = \overline{\nu}$, then we define $\chi(N\alpha) = \nu(\alpha)$ for $\alpha \in L^\times$. If $\beta \in L^\times$ is such that $N\alpha = N\beta$, then by Corollary 12.2 there exists a $\gamma \in L^\times$ such that $\alpha\beta^{-1} = \gamma\overline{\gamma}^{-1}$; hence $\nu(\alpha) = \nu(\beta)$, and therefore $\chi(N\alpha)$ is well defined. The fact that N is surjective now extends the domain of definition of χ to $K^\times$. Hence χ is a character of $K^\times$ and ν is therefore decomposable. //

Lemma 12.4: If ν is a non-decomposable character of $L^\times$, then $\sum_{Nx=\alpha} \nu(x) = 0$ for every $\alpha \in K^\times$.

Proof: By the proof of Lemma 12.3 there exists a $\lambda \in L^\times$ such that $N\lambda = 1$ and $\nu(\lambda) \neq 1$. Hence

$$\sum_{Nx=\alpha} \nu(x) = \sum_{Nx=\alpha} \nu(\lambda x) = \nu(\lambda) \sum_{Nx=\alpha} \nu(x) \quad ,$$

and our claim follows. //

We shall need the analogue to Lemma 12.3 for the trace function $Tr: L^+ \to K^+$ defined by $Tr\ x = x+\bar{x}$.

Lemma 12.5: Tr is surjective.

Proof: The trace function is a homomorphism. Its kernel consists of those x in L that satisfy $x+x^q = 0$. It contains therefore q elements. Therefore Tr is surjective. //

Corollary 12.6: If $\alpha \in L^\times$, then for every $\beta \in K$ there exists an $x \in L$ such that $\alpha x+\bar{\alpha}\bar{x} = \beta$.

13. Assigning cuspidal representations to non-decomposable characters.

Let ν be a non-decomposable character. We are going to define a representation $\rho = \rho_0$ of G that will turn to be a cuspidal representation. In order to define ρ on G, it suffices by Proposition 11.1 to define ρ as a map from $B\cup\{w'\}$ into the automorphism group of an appropriate vector space V such that the restriction of ρ to B is a homomorphism and such that ρ preserves the relations (1), (2) and (3) of section 11. The dimension of ρ should be $q-1$ by Proposition 10.2. Hence it is convenient to take V as the vector space of all functions $f: K^\times \to \mathbb{C}$.

The definition of $\mathrm{Res}_P\rho$ is motivated by the fact proved in Proposition 10.2 that it should be equal to π. Identifying the subgroup A of P with $K^\times$ and using the identity

$$\begin{pmatrix} 1 & \beta \\ 0 & 1 \end{pmatrix}\begin{pmatrix} \alpha & 0 \\ 0 & 1 \end{pmatrix} = \begin{pmatrix} \alpha & \beta \\ 0 & 1 \end{pmatrix},$$

we are led to the following definition:

$$(1) \qquad \left(\rho\begin{pmatrix} \alpha & \beta \\ 0 & 1 \end{pmatrix} f\right)(x) = \psi(\beta x) f(\alpha x) \quad .$$

Further, we would like to define that ρ coincides with ν on D:

$$(2) \qquad \left(\rho\begin{pmatrix} \delta & 0 \\ 0 & \delta \end{pmatrix} f\right)(x) = \nu(\delta) f(x) \quad .$$

It follows that we must define ρ on B by

$$(3) \qquad \left(\rho\begin{pmatrix} \alpha & \beta \\ 0 & \delta \end{pmatrix} f\right)(x) = \nu(\delta)\psi(\beta\delta^{-1}x) f(\alpha\delta^{-1}x) \quad .$$

A straightforward calculation shows that ρ is indeed a homomorphism of B into Aut V.

In order to define $\rho(w')$, we define a function $j\colon K^{\times} \to \mathbb{C}$ by

$$(4) \qquad j(u) = \frac{1}{q} \sum_{\substack{Nt=u \\ t\in L^{\times}}} \psi(t+\bar{t})\nu(t)$$

and prove that it satisfies the following two identities:

$$(5) \qquad \sum_{v\in K^{\times}} j(uv) j(v) \nu(v^{-1}) = \begin{cases} \nu(-1) & \text{if } u = 1 \\ 0 & \text{if } u \neq 1 \end{cases},$$

$$(6) \qquad \sum_{v\in K^{\times}} j(xv) j(yv) \nu(v^{-1}) \psi(v) = \nu(-1)\psi(-x-y) j(xy) \quad .$$

In these identities the free variables belong to $K^{\times}$.

<u>Proof of identity (1)</u>: We start from the left-hand side:

$$\sum_{v\in K^\times} j(uv)j(v)\nu(v^{-1}) = \sum_{v\in K^\times} q^{-2} \sum_{\substack{Nt=uv\\ Ns=v}} \psi(t+\bar{t}+s+\bar{s})\nu(ts)\nu(v^{-1}) =$$

(7)
$$= q^{-2} \sum_{v\in K^\times} \sum_{\substack{Nt=uNs\\ Ns=v}} \psi(t+\bar{t}+s+\bar{s})\nu(tsNs^{-1}) =$$

$$= q^{-2} \sum_{s\in L^\times} \sum_{Nt=uNs} \psi(t+\bar{t}+s+\bar{s})\nu(t\bar{s}^{-1}) .$$

Let $\lambda = t\bar{s}^{-1}$. Then substituting $t = \bar{s}\lambda$ in (7) we have

(8)
$$= q^{-2} \sum_{s\in L^\times} \sum_{N\lambda=u} \psi(s(1+\lambda)+\bar{s}(1+\bar{\lambda}))\nu(\lambda)$$

$$= q^{-2} \sum_{N\lambda=u} \nu(\lambda) \sum_{s\in L^\times} \psi(s(1+\lambda)+\bar{s}(1+\bar{\lambda})) .$$

For a fixed λ, the function $\psi(s(1+\lambda)+\bar{s}(1+\bar{\lambda}))$ is a character of L^+. If $\lambda \neq -1$, then this is not the unit character, since by Corollary 12.6 the map $s \mapsto s(1+\lambda)+\bar{s}(1+\bar{\lambda})$ maps L onto K and ψ is not the unit character. It follows that

$$\sum_{s\in L^\times} \psi(s(1+\lambda)+\bar{s}(1+\bar{\lambda})) = -1 .$$

If $\lambda = -1$, we have

$$\sum_{s\in L^\times} \psi(s(1+\lambda)+\bar{s}(1+\bar{\lambda})) = q^2-1 .$$

We now distinguish between two cases and suppose first that $u = 1$. Then (8) is equal to

$$= q^{-2} \sum_{\substack{N\lambda=1\\ \lambda\neq-1}} \nu(\lambda) \sum_{s\in L^\times} \psi(s(1+\lambda)+\bar{s}(1+\bar{\lambda}))+q^{-2}\nu(-1)(q^2-1)$$

$$= q^{-2} \sum_{\substack{N\lambda=1\\ \lambda\neq-1}} \nu(\lambda)+\nu(-1)q^{-2}(q^2-1) .$$

Using Lemma 12.4 we may continue the chain of equalities by

$$= q^{-2}\nu(-1)+\nu(-1)q^{-2}(q^2-1) = \nu(-1) ,$$

as desired.

Now suppose that $u \neq 1$. Then $N\lambda = u$ implies $\lambda \neq 1$. Hence (8) is equal in this case to $-q^{-2} \sum_{N\lambda=u} \nu(\lambda) = 0$ by Lemma 12.4, as desired.

Proof of identity (2): We start again from the left-hand side. Let $x,y \in K^\times$; then

$$\sum_{v\in K^\times} j(xv)j(yv)\nu(v^{-1})\psi(v) =$$

$$= q^{-2} \sum_{v\in K^\times} \sum_{\substack{Nt=xv\\ Ns=yv}} \psi(t+\bar{t}+s+\bar{s})\nu(ts)\nu(v^{-1})\psi(v) = \tag{9}$$

$$= q^{-2} \sum_{v\in K^\times} \sum_{\substack{Nt=xv\\ Ns=yv}} \psi(t+\bar{t}+s+\bar{s})\nu(tsv^{-1}) .$$

The condition $Ns = yv$ implies $tsv^{-1} = yt\bar{s}^{-1}$. Define therefore $\lambda = yt\bar{s}^{-1}$. In addition, $Nt = xv$ implies $N\lambda = xy$. Also

$$t+\bar{t}+s+\bar{s}+v = y^{-1}(s+y+\lambda)(\overline{s+y+\lambda})-y(1+y^{-1}\lambda)(\overline{1+y^{-1}\lambda}).$$

Substituting all this in (9) gives

$$= q^{-2} \sum_{v\in K^\times} \sum_{\substack{Ns=yv\\ N\lambda=xy}} \psi(y^{-1}(s+y+\lambda)(\overline{s+y+\lambda}) - y(1+y^{-1}\lambda)(\overline{1+y^{-1}\lambda}))\nu(\lambda)$$

$$= q^{-2} \sum_{s\in L^\times} \sum_{N\lambda=xy} \psi(y^{-1}(s+y+\lambda)(\overline{s+y+\lambda}) - y(1+y^{-1}\lambda)(\overline{1+y^{-1}\lambda}))\nu(\lambda) \tag{10}$$

$$= q^{-2} \sum_{N\lambda=xy} \psi(-y(1+y^{-1}\lambda)(1+y^{-1}\lambda)\nu(\lambda) \sum_{s\in L^\times}\psi(y^{-1}(s+y+\lambda)(\overline{s+y+\lambda})) .$$

Let us develop the inner sums

$$\sum_{s\in L^\times}\psi(y^{-1}(s+y+\lambda)(\overline{s+y+\lambda}) = \sum_{\substack{r\in L\\ r\neq y+\lambda}} \psi(y^{-1}r\bar{r})$$

$$= \sum_{r\in L^\times}\psi(y^{-1}r\bar{r}) + 1 - \psi(y^{-1}(y+\lambda)(\overline{y+\lambda}))$$

$$= (q+1) \sum_{u\in K^\times}\psi(y^{-1}u) + 1 - \psi(y^{-1}(y+\lambda)(y+\bar{\lambda}))$$

$$= -(q+1) + 1 - \psi(y^{-1}(y+\lambda)(y+\bar{\lambda})) .$$

We have used the fact that Ker N consists of q+1 elements. Substituting this in (10), we obtain that (10) is equal to

$$q^{-2}\sum_{N\lambda=xy}\psi(-y(1+y^{-1}\lambda)(1+y^{-1}\bar{\lambda}))\nu(\lambda)(-q-\psi(y^{-1}(y+\lambda)(y+\bar{\lambda})) =$$

$$(11)\qquad = q^{-1}\sum_{N\lambda=xy}\psi(-y(1+y^{-1}\lambda)(1+y^{-1}\bar{\lambda}))\nu(\lambda)$$

$$- q^{-2}\sum_{N\lambda=xy}\psi(-y(1+y^{-1}\lambda)(1+y^{-1}\bar{\lambda})) + y^{-1}(y+\lambda)(y+\bar{\lambda})\nu(\lambda) \quad .$$

In order to compute the two sums, note that under the assumption $\lambda\bar{\lambda} = xy$ we have

$$-y(1+y^{-1}\lambda)(1+y^{-1}\bar{\lambda}) = -y-x-(\lambda+\bar{\lambda})$$

and

$$-y(1+y^{-1}\lambda)(1+y^{-1}\bar{\lambda}) + y^{-1}(y+\lambda)(y+\bar{\lambda}) = 0 \quad .$$

Hence (7) is equal to

$$-q^{-1}\psi(-x-y)\sum_{N\lambda=xy}\psi(-\lambda-\bar{\lambda})\nu(\lambda) - q^{-2}\sum_{N\lambda=xy}\nu(\lambda) =$$

$$= -q^{-1}\psi(-x-y)\nu(-1)\sum_{N\lambda=xy}\psi(\lambda+\bar{\lambda})\nu(\lambda) = \nu(-1)\psi(-x-y)j(xy) \quad ,$$

as desired.

Having proved the above identities, we consider an $f \in V$ and define $\rho(w')f$ by

$$(12)\qquad (\rho(w')f)(y) = \sum_{x\in K^\times}\nu(x^{-1})j(yx)f(x) \quad .$$

Our task now is to prove that this definition of $\rho(w')$ together with (3) is compatible with the identities (1), (2) and (3) of section 11.

Indeed, it is convenient to write (1) of section 11 in the form

$$w'\begin{pmatrix}\alpha & 0\\ 0 & \delta\end{pmatrix} = \begin{pmatrix}\delta & 0\\ 0 & \alpha\end{pmatrix}w' \quad .$$

A straightforward calculation shows that the two automorphisms obtained by acting with ρ on both sides of the equation operate in the same way on every element $f \in V$.

In order to show that ρ preserves identity (2) of section 11, we compute for an $f \in V$:

$$(\rho(w')\rho(w')f)(z) = \sum_{x\in K^\times} \nu(x^{-1})j(zx)(\rho(w')f)(x) =$$

(13)
$$= \sum_{x\in K^\times} \nu(x^{-1})j(zx) \sum_{y\in K^\times} \nu(y^{-1})j(xy)f(y) =$$

$$= \sum_{y\in K^\times} \nu(y^{-1})f(y) \sum_{x\in K^\times} j(xy)j(zx)\nu(x^{-1}) .$$

Changing variables in (9) by $zx = v$ and $xy = uv$, (13) is equal to

$$\sum_{u\in K^\times} \nu(y^{-1})f(uz) \sum_{v\in K^\times} j(uv)j(v)\nu(v^{-1}) = \nu(-1)f(z) = \nu(-1)(\rho\begin{pmatrix} -1 & 0 \\ 0 & -1 \end{pmatrix}f)(z)$$

by (3) and by (5)

Finally we have to prove that ρ preserves relation (3) of section 11. This is done by rewriting it as

$$w'\begin{pmatrix} 1 & 1 \\ 0 & 1 \end{pmatrix}w' = \begin{pmatrix} -1 & 1 \\ 0 & -1 \end{pmatrix}w'\begin{pmatrix} 1 & -1 \\ 0 & 1 \end{pmatrix} ,$$

making the necessary computations as above, and using identity (6). The actual computations raise no significant problem, so we omit them.

We have thus proved that starting from a non-decomposable character ν of $L^\times$, there exists a representation $\rho = \rho_\nu$ of G that acts on B via (3) and on w' via (12). For later references let us also describe the action of ρ on the element

(14)
$$g = \begin{pmatrix} \alpha & \beta \\ \gamma & \delta \end{pmatrix}$$

of G, where $\gamma \neq 0$. We use the identity

$$\begin{pmatrix}\alpha & \beta\\ \gamma & \delta\end{pmatrix} = \begin{pmatrix}\beta-\alpha\gamma^{-1}\delta & -\alpha\\ 0 & -\gamma\end{pmatrix}\begin{pmatrix}0 & 1\\ -1 & 0\end{pmatrix}\begin{pmatrix}1 & \gamma^{-1}\delta\\ 0 & 1\end{pmatrix}$$

and apply (3) and (12) to compute the action of $\rho(g)$ on a function $f\colon K^\times \to \mathbb{C}$:

$$(\rho(g)f)(y) = \nu(-\gamma)\psi(\alpha\gamma^{-1}y)(\rho(w')\rho\begin{pmatrix}1 & \gamma^{-1}\delta\\ 0 & 1\end{pmatrix}f)((\beta-\alpha\gamma^{-1}\delta)(-\gamma)^{-1}y) =$$

$$= \nu(-\gamma)\psi(\alpha\gamma^{-1}y)\sum_{x\in K^\times}\nu(x^{-1})j((\alpha\gamma^{-1}\delta-\beta)\gamma^{-1}yx)(\rho\begin{pmatrix}1 & \gamma^{-1}\delta\\ 0 & 1\end{pmatrix}f)(x) =$$

$$= \nu(-\gamma)\psi(\alpha\gamma^{-1}y)\sum_{x\in K^\times}\nu(x^{-1})j(\gamma^{-2}ux\cdot\det(g))\psi(\gamma^{-1}\delta x)f(x) =$$

$$= \nu(-\gamma)\psi(\alpha\gamma^{-1}y)\sum_{x\in K^\times}\nu(x^{-1})\psi(\gamma^{-1}\delta x)f(x)(-q^{-1})\sum_{\substack{t\bar{t}=\gamma^{-2}yx\cdot\det(g)\\ t\in L^\times}}\psi(t+\bar{t})\nu(t) \quad .$$

Substitute $t = -\gamma^{-1}xu$

$$= \sum_{x\in K}\left[-q^{-1}\psi\left(\frac{ay+\delta x}{\gamma}\right)\sum_{\substack{u\bar{u}=yx^{-1}\det(g)\\ u\in L^\times}}\psi(-\tfrac{x}{\gamma}(u+\bar{u}))\nu(u)\right]f(x) \quad .$$

We have therefore proved that for an element g given by (14),

(15) $$(\rho(g)f)(y) = \sum_{x\in K^\times}k(y,x;g)f(x) \quad ,$$

where

(16) $$k(y,x;g) = \frac{1}{q}\psi\left(\frac{\alpha y+\delta x}{\gamma}\right)\sum_{\substack{u\bar{u}=yx^{-1}\det(g)\\ u\in L^\times}}\psi(-\tfrac{x}{\gamma}(u+\bar{u}))\nu(u) \quad .$$

14. The correspondence between ν and ρ_ν.

Proposition 14.1:

(a) If ν is a non-decomposable character of $L^\times$, then the representation ρ_ν of G defined in section 13 is cuspidal.

(b) If ν and ν' are non-decomposable characters of $L^\times$, then ρ_ν is isomorphic to $\rho_{\nu'}$ if and only if ν is conjugate to ν' over K.

Proof:

(a) We have defined ρ_ν such that its restriction to P is equal to π. Hence ρ_ν is cuspidal by Proposition 10.2.

(b) Let $\rho_\nu = \rho$, $\rho_{\nu'} = \rho'$, $j_\nu = j$ and $j_{\nu'} = j'$. If ν' is conjugate to ν, then j' is equal to j, as follows from the definition (4) of section 13. Hence $\rho = \rho'$.

Conversely, suppose that ρ' is isomorphic to ρ. Then there exists an automorphism θ of V (= the vector space of all functions $f\colon K^\times \to \mathbb{C}$) such that

$$\rho'(g) = \theta\rho(g)\theta^{-1} \qquad g \in G . \tag{1}$$

In particular (1) is valid for every $g \in P$. However, $\mathrm{Res}_P\rho = \pi = \mathrm{Res}_P\rho'$ and π is irreducible. Hence, by Schur's lemma, θ is a multiplication by a scalar. In particular θ commutes with every automorphism of V. Hence (1) implies that $\rho'(g) = \rho(g)$ for every $g \in G$.

It follows that for every $\delta, y \in K^\times$ and every $f \in V$

$$\nu'(\delta)f(\delta^{-1}y) = \left(\rho'\begin{pmatrix}1 & 0\\ 0 & \delta\end{pmatrix}f\right)(y) = \left(\rho\begin{pmatrix}1 & 0\\ 0 & \delta\end{pmatrix}\right)f(y) = \nu(\delta)f(\delta^{-1}y) .$$

Hence

$$\nu'(\delta) = \nu(\delta) \quad \text{for every } \delta \in K^\times . \tag{2}$$

Further $\rho'(w') = \rho(w')$; hence

$$\sum_{x\in K^\times}\nu'(x^{-1})j(yx)f(x) = \sum_{x\in K^\times}\nu(x^{-1})j'(yx)f(x)$$

for every $y \in K^{\times}$ and every $f \in V$. Using (2) this implies that $j'(u) = j(u)$ for every $u \in K$. This means that

$$(3) \qquad \sum_{Nt=u} \psi(t+\bar{t})\nu'(t) = \sum_{Nt=u} \psi(t+\bar{t})\nu(t) \quad .$$

Let $\delta \in K^{\times}$ and replace the variable t in (3) by δt. Using (2) to cancel $\nu(\delta)$ on both sides, we have

$$(4) \qquad \sum_{t\bar{t}=v} \psi(\delta(t+\bar{t}))\nu'(t) = \sum_{t\bar{t}=v} \psi(\delta(t+\bar{t}))\nu(t)$$

for every $v,\delta \in K^{\times}$. This can be rewritten as

$$(5) \qquad \sum_{t\bar{t}=v}{}' (\nu'(t)+\nu'(\bar{t})-\nu(t)-\nu(\bar{t}))\psi(\delta(t+\bar{t})) = 0 \quad ,$$

where the prime over the summation symbol indicates that for every $t \in L^{\times}$ such that $t\bar{t} = v$, only one pair out of the two $(t,\bar{t})$, $(\bar{t},t)$ contributes a summand to the summation.

Now let x be an element of K. Then there exists a $t \in L$ such that $t+\bar{t} = x$ and $t\bar{t} = v$. This element is a solution to the quadratic equation $X^2-xX+v = 0$. (We repeat that L is the unique quadratic extention of K.) The other solution of the equation is obviously $\bar{t}$. The expression $a_x = \nu'(t)+\nu'(\bar{t})-\nu(t)-\nu(\bar{t})$ is therefore well defined and (5) can be rewritten as

$$\sum_{t\bar{t}=v} a_x\psi(\delta x) = 0 \quad .$$

If $x_1 \neq x_2$, then the characters $\delta \mapsto \psi(\delta x_1)$ and $\delta \mapsto \psi(\delta x_2)$ of K^+ are distinct. Hence by Artin's lemma, $a_x = 0$ for every $x \in K$. This means that

$$\nu'(t) + \nu'(\bar{t}) = \nu(t) + \nu(\bar{t}) \qquad t \in L^{\times} \quad .$$

In addition,

$$\nu'(t)\nu'(\bar{t}) = \nu(t)\nu(\bar{t}) \qquad t \in L^{\times}$$

by (2). Hence the pairs $(\nu'(t),\nu'(\bar{t}))$, $(\nu(t),\nu(\bar{t}))$ are the solutions of the same quadratic equation over K. Hence

(6) $\{\nu'(t),\nu'(\bar{t})\} = \{\nu(t),\nu(\bar{t})\} \qquad t \in L^{\times}$.

In particular, (6) is true for a generator t_0 of the cyclic group $L^{\times}$. Suppose, for example, that $\nu(t_0) = \nu'(t_0)$. Then $\nu(t) = \nu'(t)$ for every $t \in L^{\times}$. If $\nu'(t_0) = \nu(\bar{t}_0)$, then $\nu' = \bar{\nu}$.

We have therefore proved that ν' is conjugate to ν over K. //

At this point we would like to indicate an interesting duality between conjugacy classes of G on one hand and irreducible representations of G on the other hand. The elements α of $K^{\times}$ correspond bijectively to the pair of conjugacy classes $(c_1(\alpha),\ c_2(\alpha))$ of G (see section 5), and there are $q-1$ of them. Dually the characters μ_1 of $K^{\times}$ correspond bijectively to pairs $(\rho_{(\mu_1,\mu_1)},\ \rho'_{(\mu_1,\mu_1)})$ of irreducible representations of dimensions q and 1, respectively (see Theorem 8.12), and there are $q-1$ of them. Further, the pairs of elements α,β of $K^{\times}$ with $\alpha \neq \beta$ correspond to the conjugacy classes $c_3(\alpha,\beta)$, and there are $\frac{1}{2}(q-1)(q-2)$ of them. Dually the pairs of characters μ_1,μ_2 of K with $\mu_1 \neq \mu_2$ correspond to the irreducible representations $\rho_{(\mu_1,\mu_2)}$ of G of dimension $q+1$, and there are also $\frac{1}{2}(q-1)(q-2)$ of them. Finally, the elements λ of $L^{\times}-K^{\times}$ correspond to the conjugacy classes $c_4(\lambda)$, whereas the characters ν of $L^{\times}$ that do not come from characters of $K^{\times}$ (i.e., non-decomposable) correspond to the cuspidal representations ρ_ν of G. In both sets there are $\frac{1}{2}(q^2-q)$ elements.

We summarize these data in the following table.

elements of $L^\times$	conjugacy classes	characters of $L^\times, K^\times$	irr. repr. of G	dim of repr.	no. of elements
$\alpha \in K^\times$	$c_1(\alpha)$	$\mu_1 \in X(K^\times)$	$\rho'(\mu_1,\mu_1)$	1	q-1
	$c_2(\alpha)$		$\rho(\mu_1,\mu_1)$	q	q-1
$\alpha, \beta \in K^\times$ $\alpha \neq \beta$	$c_3(\alpha,\beta)$	$\mu_1, \mu_2 \in X(K^\times)$ $\mu_1 \neq \mu_2$	$\rho(\mu_1,\mu_1)$	q+1	$\frac{1}{2}(q-1)(q-2)$
$\lambda \in L^\times - K^\times$	$c_4(\lambda)$	$\nu \in X(L^\times) - X(K^\times)$	ρ_ν	q-1	$\frac{1}{2}(q^2-q)$

15. The small Weil group and the reciprocity law.

Let F/E be a finite Galois extension. Its Galois group, $G(F/E)$, acts on the multiplicative group $F^\times$ of F. Denote by $W(F/E) = G(F/E)\cdot F^\times$ the semi-direct product of $G(F/E)$ by $F^\times$. It consists of all pairs (x,σ) where $x \in F^\times$ and $\sigma \in G(F/E)$. Multiplication is given by the formula

$$(x,\sigma)\cdot(y,\tau) = (x\cdot\sigma y, \sigma\tau) \quad ;$$

the one element is (1,1) and the inverse is given by

$$(x,\sigma)^{-1} = (\sigma^{-1}x^{-1}, \sigma^{-1}) \quad .$$

The map $x \mapsto (x,1)$ is an embedding of $F^\times$ in $W(F/E)$. We identify $F^\times$ with its image. $F^\times$ is normal in $W(F/E)$ and its index is equal to the degree $[F:E]$.

The group $W(F/E)$ is in general not abelian. A typical commutator is

$$(x,\sigma)(y,\tau)(x,\sigma)^{-1}(y,\tau)^{-1} = (x\cdot\sigma(y)\cdot\sigma\tau\sigma^{-1}(x^{-1})\cdot\sigma\tau\sigma^{-1}\tau^{-1}(y), \sigma\tau\sigma^{-1}\tau^{-1}).$$

If in addition $G(F/E)$ is abelian, then this formula simplifies to

$$(1) \qquad (x,\sigma)(y,\tau)(x,\sigma)^{-1}(y,\tau)^{-1} = (x\cdot\sigma(y)\cdot\tau(x^{-1})\cdot y,\ 1) \quad .$$

We now restrict our attention to the case where $E = K$ is our field with q elements and $F = L$ is its unique quadratic extension. The Galois group $G(L/K)$ consists of two elements, a conjugation, the action of which is denoted by a bar, and the identity automorphism. In this case $W(L/K)$ is called the small Weil group of the extension L/K. It is a finite group having $2(q-1)$ elements, and it can be described as the free group generated by $L^\times$ and with the relations

(2) $\quad \varphi^2 = 1 \quad$ and $\quad x\varphi = \varphi\bar{x} \quad$ for $x \in L^\times$.

Now $W(L/K)$ contains the abelian normal subgroup $L^\times$ of index 2. Hence its irreducible representations are of degree ≤ 2.

We would like to establish a correspondence between the two-dimensional representations of $W(L/K)$ (not only the irreducible ones) and the higher dimensional representations of G. As a first step toward this goal, let us compute the number of characters of $W(L/K)$. This number is equal to the index of the commutator subgroup $W(L/K)'$ of $W(L/K)$. Indeed, applying formula (1) to the four possible pairs (σ,τ), one concludes that

(3) $\quad W(L/K)' = \{z\bar{z}^{-1} \mid z \in L^\times\}$.

By Corollary 12.2 the right-hand side is equal to $\{x \in L^\times \mid Nx = 1\}$; hence $(W(L/K):W(L/K)') = 2(q-1)$ by Lemma 12.1. We have therefore proved:

<u>Lemma 15.1</u>: $W(L/K)$ has $2(q-1)$ characters.

Consider now a two-dimensional representation τ of $W(L/K)$. Its restriction to $L^\times$ decomposes into a direct sum of two characters. Let ν be one of them.

<u>Lemma 15.2</u>:

(a) If ν is non-decomposable, then $\mathrm{Res}_{L^\times}\tau = \nu \oplus \bar{\nu}$.

(b) ν is non-decomposable if and only if τ is irreducible.

Proof: By construction there exists a vector $0 \neq v_1 \in V_\tau$ such that

(4) $\tau(x)v_1 = \nu(x)v_1 x \in L^\times$.

Let

(5) $v_1' = \tau(\varphi)v_1$.

Then the relation $x\varphi = \varphi\bar{x}$ implies

(6) $\tau(x)v_1' = \nu(\bar{x})v_1' \quad x \in L^\times$.

Hence $\bar{\nu}$ is also a component of $\mathrm{Res}_{L^\times}\tau$. If $\nu \neq \bar{\nu}$, then $\mathrm{Res}_{L^\times}\tau = \nu \oplus \bar{\nu}$, and (a) is proved.

In order to prove (b), suppose first that $\tau = \tau_1 \oplus \tau_2$ is reducible. Then τ_i are characters of $W(L/K)$ and since $x\bar{x}^{-1} \in W(L/K)'$, by (3) we have $\tau_i(x) = \tau_i(\bar{x})$. It follows that $\mathrm{Res}_{L^\times}\tau_i$ are decomposable characters of $L^\times$. Also ν must be equal to one of them; hence ν is also decomposable.

Conversely, suppose that $\nu = \bar{\nu}$. There are two possibilities. Either $v_1' = \nu(\varphi)v_1$ is a multiple of v_1, or v_1 and v_1' are linearly independent. In the first case $\nu: L^\times \cup \{\varphi\} \to \mathbb{C}$ is a map which is multiplicative on $L^\times$ and agrees with the defining relations (3) of $W(L/K)$. Hence ν can be extended to a character of $W(L/K)$ that happens to be a component of τ by (4) and (5). This implies that τ is reducible.

In the second case v_1 and v_1' generate V_τ. Then (4) and (6) imply that $\tau(x)v = \nu(x)v$ for every $x \in L^\times$ and $v \in V_\tau$. Let v_2 be an eigenvector of $\tau(\varphi)$. Then since $\tau(x)v_2 = \nu(x)v_2$, we prove as above that v_2 is an eigenvector of $W(L/K)$. This implies that τ is reducible. //

Now consider more closely the case where $\tau = \tau_1 \oplus \tau_2$ is a reducible representation of $W(L/K)$. By Lemma 15.2 there exist characters μ_1 and μ_2 of $K^\times$ such that $\tau_i(x) = \mu_i(Nx)$ for $i = 1,2$ and for every $x \in L^\times$. We may therefore write $\tau = \tau_{(\mu_1,\mu_2)}$. The pair (μ_1,μ_2) defines a character μ of

B (by (1) of section 7). If $\mu_1 = \mu_2$, then $\hat{\mu} = \rho_{(\mu_1,\mu_1)} \oplus \rho'_{(\mu_1,\mu_1)}$. We correspond τ to $\rho_{(\mu_1,\mu_1)}$. If $\mu_1 \neq \mu_2$, then $\hat{\mu} = \rho_{(\mu_1,\mu_2)}$. We correspond τ to $\rho_{(\mu_1,\mu_2)}$.

This correspondence is bijective. Indeed, starting from a pair of characters (μ_1,μ_2) of $K^\times$, we define characters τ_1,τ_2 of $W(L/K)$ by $\tau_i(x) = \mu_i(Nx)$ and $\tau_i(\varphi) = 1$. This definition makes sense, since it is compatible with the relations (2). Then $\tau_{(\mu_1,\mu_2)} = \tau_1 \oplus \tau_2$ is a reducible two-dimensional character of $W(L/K)$ and $\rho_{(\mu_1,\mu_2)}$ corresponds to $\tau_{(\mu_1,\mu_2)}$.

Finally, let ν be a non-decomposable character of $L^\times$. Define a two-dimensional representation τ_ν of $W(L/K)$ by

$$\tau_\nu(x) = \begin{pmatrix} \nu(x) & 0 \\ 0 & \nu(\bar{x}) \end{pmatrix} \quad \text{for } x \in L^\times, \text{ and } \quad \tau_\nu(\varphi) = \begin{pmatrix} 0 & 1 \\ 1 & 0 \end{pmatrix} .$$

Then $\operatorname{Res}_{L^\times} \tau_\nu = \nu \oplus \bar{\nu}$; hence τ_ν is irreducible. If ν' is an additional non-decomposable character of $L^\times$, then $\tau_\nu = \tau_{\nu'}$ if and only if ν' is conjugate to ν. It follows that $W(L/K)$ has exactly $\frac{1}{2}(q^2-q)$ irreducible representations of the form τ_ν. Taking into account Lemma 15.1, we deduce from the identity

$$2(q^2-1) = 2^2 \cdot \tfrac{1}{2}(q^2-q) + 1^2 \cdot 2(q-1)$$

that the τ_ν are all the irreducible two-dimensional representations of $W(L/K)$.

Adding Proposition 14.1 to the above arguments, we have proved the following reciprocity law.

<u>Theorem 15.3</u>: There exists a bijective correspondence between the two-dimensional representations τ of the Weil group $W(L/K)$ and the higher dimensional representations of $GL(2,K)$.

Using the above notation, the correspondence may be described as follows:

(a) For reducible τ: If (μ_1,μ_2) is a pair of characters of $K^\times$, then

$\tau_{(\mu_1,\mu_2)}$ corresponds to $\rho_{(\mu_1,\mu_2)}$;

(b) For irreducible τ: If ν is a non-decomposable character of $L^\times$, then τ_ν corresponds to ρ_ν.

Chapter 3. Bessel functions and Γ-functions

16. Whittakers models.

Recall that we have fixed a non-unit character ψ of K^+, identified it with a character of U and found that $\pi = \mathrm{Ind}_U^P \psi$ is an irreducible representation of dimension $q-1$. If χ is a character of G, then by the Frobenius reciprocity theorem $(\chi, \mathrm{Ind}_U^G \psi)_G = (\chi, \mathrm{Ind}_B^G \pi)_G = (\mathrm{Res}_B \chi, \pi)_B = 0$. Hence all irreducible components of $\mathrm{Ind}_U^G \psi$ are of dimension > 1. Each one of them appears in multiplicity 1 by Proposition 10.3. On the other hand, if we sum up the dimensions of all higher dimensional irreducible representations of G, we find that it is equal to $\dim \mathrm{Ind}_U^G \psi = (q-1)^2(q+1)$. This follows from the table in section 14 and from the following identity:

$$\tfrac{1}{2}(q^2-q)(q-1)+(q-1)q + \tfrac{1}{2}(q-1)(q-2)(q+1) = (q-1)^2(q+1) \quad .$$

We conclude therefore that

<u>Theorem 16.1</u>: $\mathrm{Ind}_U^G \psi$ is equal to the direct sum of all higher dimensional representations of G, each with multiplicity 1.

Note that the same conclusion can be drawn from the following lemma, which is proved, however, without explicit use of the finiteness of K.

<u>Lemma 16.2</u>: Let ρ be a representation of G of dimension > 1. Then:

(a) $\mathrm{res}_P V_\rho = \mathrm{Res}_P J(V_\rho) \oplus V_\pi$;

(b) $\dim J(V_p) = \dim \rho - (q-1)$;

(c) The multiplicity of ρ in $\mathrm{Ind}_U^G \psi$ is equal to 1.

<u>Proof</u>: If $\dim \rho = q-1$, i.e., if ρ is cuspidal, then $J(V_\rho) = 0$ and

$\text{Res}_P V_\rho = V_\pi$ by Proposition 10.2. Formula (a) is therefore true in this case.

If $\dim \rho \geq q$, i.e., if ρ is non-cuspidal, then $J(V_\rho) \neq 0$ and there exists a character μ of B and a representation ρ' of G such that $\hat{\mu} = \text{Ind}_B^G \mu = \rho' \oplus \rho$ and $\dim \rho' \leq 1$. Further, by Lemma 8.8 we have $\text{Res}_P V_{\hat{\mu}} = \text{Res}_P J(V_{\hat{\mu}}) \oplus V_\pi$; hence

$$(1) \qquad \text{Res}_P V_{\rho'} \oplus \text{Res}_P V_\rho = \text{Res}_P J(V_{\rho'}) \oplus \text{Res}_P J(V_\rho) \oplus V_\pi \quad .$$

Now V_π does not contain any one-dimensional P-module since $\dim V_\pi = q-1 > 1$ and V_π is P-irreducible. It follows that $V_\pi \subseteq \text{Res}_P V_\rho$. Further, $\text{Res}_P J(V_\rho)$ is certainly contained in V_ρ, and by Lemma 8.2 it decomposes into a direct sum of one-dimensional P-subspaces. Hence $\text{Res}_P V_\pi \cap \text{Res}_P J(V_\rho) = 0$, and thus the right-hand side of (a) is contained in its left-hand side. If $\dim \rho = q$, then since the dimension of the right-hand side of (a) is $\geq 1+(q-1)$, we may conclude the equality. If $\dim \rho = q+1$, then $\rho' = 0$ and (1) coincides with (a).

Note that we have also proved that the multiplicity of π in $\text{Res}_P \rho$ is 1. Hence, by the Frobenius reciprocity theorem the multiplicity of ρ in $\text{Ind}_U^G \psi = \text{Ind}_B^G \pi$ is 1. //

Recall that $\text{Ind}_U^G V_\psi$ can be identified with the space of all functions $F: G \to \mathbb{C}$ such that

$$F(ug) = \psi(u)F(g) \quad \text{for } u \in U \text{ and } g \in G \quad .$$

The group G operates on $\text{Ind}_U^G V_\psi$ by the following law:

$$(sF)(g) = F(gs).$$

If ρ is now an irreducible higher dimensional representation of G, then V_ρ can be embedded in $\text{Ind}_U^G V_\psi$. For every $v \in V_\rho$ there exists therefore a function $W_v: G \to \mathbb{C}$ called a <u>Whittaker function</u> of ρ such that the following rules hold:

$$W_v = 0 \iff v = 0 \quad ,$$

$$W_{c_1v_1+c_2v_2} = c_1W_{v_1}+c_2W_{v_2} \qquad \text{for } c_1,c_2 \in \mathbb{C} \quad ,$$

(2) $\quad W_v(ug) = \psi(u)W_v(g) \qquad$ for $u \in U$ and $g \in G$,

(3) $\quad W_{\rho(s)v}(g) = W_v(gs) \qquad$ for $s,g \in G$.

The set of all functions W_v form a G-subspace $W(\rho)$ of $\mathrm{Ind}_U^G V_\psi$ called the <u>Whittaker model</u> of ρ. By Theorem 16.1 or by Lemma 16.2, this subspace is uniquely determined within $\mathrm{Ind}_U^G V_\psi$. Moreover, if ρ' is an additional higher dimensional representation of G, then $W(\rho) \cap W(\rho') = 0$.

17. The Γ-function of a representation.

Let ρ be a higher dimensional representation of G. The P-decomposition

(1) $\quad V_\rho = J(V_\rho)\oplus V_\pi$

of V_ρ, obtained in Lemma 16.2, is also an A-decomposition, where we recall that

$$A = \left\{\begin{pmatrix}\alpha & 0\\ 0 & 1\end{pmatrix} \;\middle|\; \alpha \in K^\times\right\}$$

is the subgroup of P which is canonically isomorphic to $K^\times$. If $\dim \rho \geq q$, then $\rho = \rho_{(\mu_1,\mu_2)}$, where μ_1,μ_2 are characters of $K^\times$. By Lemma 8.2 they are the eigenvalues of A on $J(V_\rho)$. If $\dim \rho = q+1$, then $\mu_1 \neq \mu_2$ and $\dim J(V_\rho) = 2$. If $\dim \rho = q$, then $\mu_1 = \mu_2$ and $\dim J(V_\rho) = 1$. If $\dim \rho = q-1$, then $J(V_\rho) = 0$. In the first two cases we call μ_1^{-1}, μ_2^{-1} and μ_1^{-1}, respectively, the <u>exceptional characters</u> for ρ. In the third case there are no exceptional characters. In any case, if the inverse of a character ω of $K^\times$ is not exceptional, then it is not an eigenvalue of A operating on $J(V_\rho)$ through ρ.

Lemma 17.1: If a character ω of $K^\times$ is not an exceptional character for ρ, then any two linear functionals ℓ_1, ℓ_2 of V_ρ satisfying

$$\ell_i(\rho\begin{pmatrix} x & 0 \\ 0 & 1 \end{pmatrix}v) = \omega(x)^{-1}\ell_i(v) \qquad \text{for every } x \in K^\times \text{ and } v \in V_\rho$$

are linearly dependent.

Proof: $\mathrm{Res}_A V_\pi$ is isomorphic to the space of all functions $\varphi\colon K^\times \to \mathbb{C}$, and A acts on this space by the formula

$$(\rho\begin{pmatrix} \alpha & 0 \\ 0 & 1 \end{pmatrix}\varphi)(x) = \varphi(x\alpha) \quad .$$

Indeed, we map an $f \in V_\pi$ to the function φ defined by

$$\varphi(x) = f\begin{pmatrix} x & 0 \\ 0 & 1 \end{pmatrix} .$$

On the other hand the identity

$$\begin{pmatrix} 1 & \beta \\ 0 & 1 \end{pmatrix}\begin{pmatrix} \alpha & 0 \\ 0 & 1 \end{pmatrix} = \begin{pmatrix} \alpha & \beta \\ 0 & 1 \end{pmatrix}$$

implies that φ also determines f.

There exists now exactly one non-zero function φ (up to a multiplication by a scalar) such that $\varphi(x\alpha) = \omega(\alpha)^{-1}\varphi(x)$. This function is defined by $\varphi(x) = \omega(x)^{-1}\varphi(1)$. Using (1) and the assumption that the inverse ω is non-exceptional, this means that ω is an eigenvalue of A operating on V_ρ and the subspace of eigenvectors of A belonging to ω is one-dimensional.

Now let ζ be a generator of the cyclic group $K^\times$ and define a linear map $T\colon V \to V$ by

$$Tv = \rho\begin{pmatrix} \zeta & 0 \\ 0 & 1 \end{pmatrix}v - \omega(\zeta)^{-1}v \quad .$$

Then

$$\mathrm{Ker}\, T = \{v \in V_\rho \mid \rho\begin{pmatrix} x & 0 \\ 0 & 1 \end{pmatrix}v = \omega(x)^{-1}v \quad \text{for all } x \in K^\times\}$$

is the subspace of eigenvectors of A belonging to ω^{-1}. We proved that $\dim \operatorname{Ker} T = 1$. Hence $\dim T(V_\rho) = \dim \rho - 1$. However, $T(V) \subseteq \operatorname{Ker} \ell_i$ and $\dim \operatorname{Ker} \ell_i = \dim \rho - 1$. Hence $\operatorname{Ker} \ell_1 = T(V) = \operatorname{Ker} \ell_2$. It follows that ℓ_1 and ℓ_2 are linearly dependent. //

Theorem 17.2: Let ρ be a higher dimensional irreducible representation of G and let ω be a character of $K^\times$ which is not exceptional for ρ. Then there exists a complex number $\Gamma_\rho(\omega)$ such that for every Whittaker function W_v of ρ we have

$$\Gamma_\rho(\omega) \sum_{x\in K^\times} W_v\begin{pmatrix} x & 0 \\ 0 & 1 \end{pmatrix}\omega(x) = \sum_{x\in K^\times} W_v\begin{pmatrix} 0 & 1 \\ x & 0 \end{pmatrix}\omega(x) .$$

Proof: Define linear functionals ℓ_i, $i = 1,2$ of V_ρ by

$$\ell_1(v) = \sum_{x\in K^\times} W_v\begin{pmatrix} x & 0 \\ 0 & 1 \end{pmatrix}\omega(x) \qquad \text{and}$$

$$\ell_2(v) = \sum_{x\in K^\times} W_v\begin{pmatrix} 0 & 1 \\ x & 0 \end{pmatrix}\omega(x) .$$

Then

$$\ell_i(\rho\begin{pmatrix} \alpha & 0 \\ 0 & 1 \end{pmatrix}v) = \omega(\alpha)^{-1}v \qquad i = 1,2$$

for every $\alpha \in K^\times$ and every $v \in V_\rho$. For example

$$\ell_2(\rho\begin{pmatrix} \alpha & 0 \\ 0 & 1 \end{pmatrix}v) = \sum_{x\in K^\times} W_{\rho\begin{pmatrix} \alpha & 0 \\ 0 & 1 \end{pmatrix}v}\begin{pmatrix} 0 & 1 \\ x & 1 \end{pmatrix}\omega(x) = \sum_{x\in K^\times} W_v\begin{pmatrix} 0 & 1 \\ x\alpha & 0 \end{pmatrix}\omega(x) =$$

$$= \omega(\alpha)^{-1} \sum_{y\in K^\times} W_v\begin{pmatrix} 0 & 1 \\ y & 0 \end{pmatrix}\omega(y) = \omega(\alpha)^{-1}\ell_2(v) .$$

It follows from Lemma 17.1 that ℓ_2 is a multiple of ℓ_1 by a constant. We denote this constant by $\Gamma_\rho(\omega)$. //

The complex valued function $\Gamma_\rho(\omega)$ defined for every non-exceptional character ω of $K^\times$ will play an important role in the computation of the character table of G.

18. Determination of ρ by Γ_ρ.

Let ρ be a higher dimensional representation of G. For every $v \in V_\rho$, let W_v be the corresponding Whittaker function of ρ and let r be the homomorphism of V_ρ into the space $F(K^\times,\mathbb{C})$ of all functions $\varphi: K^\times \to \mathbb{C}$, defined by $r(v) = \mathrm{Res}_A W_v$. If we define an operation of $K^\times$ on $F(K^\times,\mathbb{C})$ by $(\alpha\cdot\varphi)(x) = \varphi(x\alpha)$ and identify A with $K^\times$, then r is also an A-homomorphism.

Lemma 18.1: The homomorphism r is surjective and $\mathrm{Ker}\ r = J(V_\rho)$.

Proof: We start by determining the kernel of ρ. Let $v \in J(V_\rho)$; then for every $\alpha \in K^\times$ we choose a $\beta \in K$ such that $\psi(\alpha\beta) \neq 0$. Then

$$W_v\begin{pmatrix}\alpha & 0\\ 0 & 1\end{pmatrix} = W_{\rho\begin{pmatrix}1 & \beta\\ 0 & 1\end{pmatrix}v}\begin{pmatrix}\alpha & 0\\ 0 & 1\end{pmatrix} = W_v\begin{pmatrix}\alpha & \alpha\beta\\ 0 & 1\end{pmatrix} = W_v\left[\begin{pmatrix}1 & \alpha\beta\\ 0 & 1\end{pmatrix}\begin{pmatrix}\alpha & 0\\ 0 & 1\end{pmatrix}\right] =$$

$$= \psi(\alpha\beta)W_v\begin{pmatrix}\alpha & 0\\ 0 & 1\end{pmatrix} .$$

Hence $W_v\begin{pmatrix}\alpha & 0\\ 0 & 1\end{pmatrix} = 0$, i.e., $v \in \mathrm{Ker}(r)$.

To prove that $\mathrm{Ker}\ r \subseteq J(V_\rho)$, recall again the P-decomposition $V = J(V_\rho)\oplus V_\pi$ of V (cf. Lemma 16.2). Then $V_\pi \cap \mathrm{Ker}\ r$ is left invariant by P. This follows from the decomposition $P = AU$ and from the following two computations: Let $v \in V_\pi\cap\mathrm{Ker}\ r$; then

$$W_{\rho\begin{pmatrix}\alpha & 0\\ 0 & 1\end{pmatrix}v}\begin{pmatrix}\alpha' & 0\\ 0 & 1\end{pmatrix} = W_v\begin{pmatrix}\alpha\alpha' & 0\\ 0 & 1\end{pmatrix} = 0$$

(1) $$W_{\rho\begin{pmatrix}1 & \beta\\ 0 & 1\end{pmatrix}v}\begin{pmatrix}\alpha & 1\\ 0 & 1\end{pmatrix} = \psi(\alpha\beta)W_v\begin{pmatrix}\alpha & 0\\ 0 & 1\end{pmatrix} = 0 \quad .$$

It follows that $V_\pi \cap \operatorname{Ker} r = 0$ or $V_\pi \cap \operatorname{Ker} r = V_\pi$ since V_π is P-irreducible.

Assume that $V_\pi \cap \operatorname{Ker} r = V_\pi$. Then by the first part of the proof, $W_v(a) = 0$ for every $v \in V$ and every $a \in A$. Hence, if $g \in G$, then $W_v(g) = W_{\rho(g)v}(1) = 0$, i.e., $v = 0$, which is a contradiction. It follows that $V_\pi \cap \operatorname{Ker} r = 0$; hence $\operatorname{Ker} r = J(V_\rho)$. This fact implies that $\dim \operatorname{Im} r = \dim V_\rho - \dim J(V_\rho) = q-1 = \dim F(K^\times,\mathbb{C})$. Hence $\operatorname{Im} r = F(K^\times,\mathbb{C})$. //

The center Z of G consists of the scalar matrices and is therefore canonically isomorphic to $K^\times$. The restriction of ρ to Z can therefore be identified with a character ω_ρ of $K^\times$, called the central character of ρ.

Proposition 18.2: A cuspidal representation ρ of G is uniquely determined by its Γ-function and its central character.

Proof: Let ρ and ρ' be two cuspidal representations of G. Then by Proposition 10.2 $\operatorname{Res}_P\rho = \pi = \operatorname{Res}_P\rho'$. If ρ and ρ' coincide on Z, then they coincide on B since $B = ZP$. Suppose in addition that $\Gamma_\rho = \Gamma_{\rho'}$. We have to prove that $\rho = \rho'$. The Bruhat decomposition $G = B \cup BwU$ implies that it suffices to show that $\rho(w) = \rho'(w)$.

Both representations ρ and ρ' are of dimension $q-1$. We can therefore assume that both of them act on the same space V. For every $v \in V$ let W_v and W'_v be Whittaker functions of ρ and ρ', respectively. We know that $J(V_\rho) = J(V_{\rho'}) = 0$; hence by Lemma 18.1 the maps $v \mapsto \operatorname{Res}_A W_v$ and $v \mapsto \operatorname{Res}_A W'_v$ are A-isomorphisms of V onto $F(K^\times,\mathbb{C})$. Hence, without loss of generality, we can assume that $\operatorname{Res}_A W_v = \operatorname{Res}_A W'_v$.

Therefore, by Theorem 17.2 the assumption $\Gamma_\rho = \rho'$ implies that for every character ω of $K^\times$,

$$\sum_{x\in K^\times} W_v\begin{pmatrix}0 & 1\\ x & 0\end{pmatrix}\omega(x) = \sum_{x\in K^\times} W'_v\begin{pmatrix}0 & 1\\ x & 0\end{pmatrix}\omega(x) \quad .$$

This implies by Artin's lemma that

$$W_v\begin{pmatrix}0 & 1\\ x & 0\end{pmatrix} = W'_v\begin{pmatrix}0 & 1\\ x & 0\end{pmatrix} \qquad \text{for every} \quad x \in K^\times .$$

Hence

$$W_{\rho(w)v}\begin{pmatrix}x & 0\\ 0 & 1\end{pmatrix} = W_v\begin{pmatrix}x & 0\\ 0 & 1\end{pmatrix}\begin{pmatrix}0 & 1\\ 1 & 0\end{pmatrix} = W_v\begin{pmatrix}0 & 1\\ x^{-1} & 0\end{pmatrix}\begin{pmatrix}x & 0\\ 0 & x\end{pmatrix} = W_{\rho\begin{pmatrix}x & 0\\ 0 & x\end{pmatrix}v}\begin{pmatrix}0 & 1\\ x^{-1} & 0\end{pmatrix}$$

$$= W_{\rho'\begin{pmatrix}x & 0\\ 0 & x\end{pmatrix}v}\begin{pmatrix}0 & 1\\ x^{-1} & 0\end{pmatrix} = W'_{\rho'\begin{pmatrix}x & 0\\ 0 & x\end{pmatrix}v}\begin{pmatrix}0 & 1\\ x^{-1} & 0\end{pmatrix} =$$

$$= W'_{\rho'(w)v}\begin{pmatrix}x & 0\\ 0 & 1\end{pmatrix} .$$

Hence by Lemma 18.1, $W_{\rho(x)v} = W'_{\rho'(w)v}$, and hence $\rho(w) = \rho'(w)$. //

19. The Bessel function of a representation.

Let ρ be a higher dimensional representation of G. Then $\dim V_\rho > 2 \geqslant \dim J(V_\rho)$ except in the case where $q = 3$ and $\dim \rho = 2$. In this case ρ is however cuspidal and $J(V_\rho) = 0$. Therefore $J(V_\rho) \neq V_\rho$ in all cases.

As U is an abelian group, $\mathrm{Res}_U \rho$ decomposes into a direct sum of characters. One of them must be different from the unit character. Indeed, otherwise we would have that

$$\rho\begin{pmatrix}1 & \beta\\ 0 & 1\end{pmatrix}v = v$$

for every $\beta \in K$ and every $v \in V_\rho$. Then by (1) of section 18,

$$W_v\begin{pmatrix}\alpha & U\\ 0 & 1\end{pmatrix} = \psi(\alpha\beta)W_v\begin{pmatrix}\alpha & 0\\ 0 & 1\end{pmatrix}$$

for every $\alpha \in K$. Hence $\mathrm{Res}_A W_v = 0$, and hence $v \in J(V_\rho)$ by Lemma 18.1. This contradicts the inequality $J(V_\rho) \neq V_\rho$. There exists therefore a non-unit character ψ_1 of K^+ and a non-zero vector $v'_1 \in V_\rho$ such that

$$\rho\begin{pmatrix}1 & \beta\\ 0 & 1\end{pmatrix}v_1' = \psi_1(\beta)v_1 \quad \text{for every} \quad \beta \in K \ .$$

By section 6 there exists an $\alpha \in K^\times$ such that $\psi_1(\beta) = \psi(\alpha\beta)$. Using the identity

$$\begin{pmatrix}\alpha & 0\\ 0 & 1\end{pmatrix}\begin{pmatrix}1 & \alpha^{-1}y\\ 0 & 1\end{pmatrix}\begin{pmatrix}\alpha^{-1} & 0\\ 0 & 1\end{pmatrix} = \begin{pmatrix}1 & y\\ 0 & 1\end{pmatrix}$$

and replacing v_1' by $v_1 = \rho\begin{pmatrix}\alpha & 0\\ 0 & 1\end{pmatrix}v_1'$, we get that

(1) $$\rho\begin{pmatrix}1 & \beta\\ 0 & 1\end{pmatrix}v_1 = \psi(\beta)v_1 \quad \text{for every} \quad \beta \in K \ .$$

It follows that if $\alpha \in K^\times$, then

$$\psi(\beta)W_{v_1}\begin{pmatrix}\alpha & 0\\ 0 & 1\end{pmatrix} = \psi(\alpha\beta)W_{v_1}\begin{pmatrix}\alpha & 0\\ 0 & 1\end{pmatrix} \quad \text{for all} \quad \beta \in K \ .$$

If $\alpha \neq 1$, then we may conclude that

(2) $$W_{v_1}\begin{pmatrix}\alpha & 0\\ 0 & 1\end{pmatrix} = 0 \ .$$

Further, (1) implies that $v_1 \notin J(V_\rho)$. Hence $\mathrm{Res}_A W_{v_1} \neq 0$ by Lemma 18.1 and therefore $W_{v_1}(1) \neq 0$. The vector v_1 is said to be a <u>Bessel vector</u> for ρ.

If v_2 is an additional Bessel vector for ρ, then by the last paragraph there exists a $\zeta \in \mathbb{C}$ such that $W_{v_1}(a) = \zeta W_{v_2}(a)$ for every $a \in A$. Using Lemma 18.1 once again, we conclude that $v_1 - \zeta v_2 \in J(V_\rho)$. Hence

$$\psi(\beta)(v_1-\zeta v_2) = \rho\begin{pmatrix}1 & \beta\\ 0 & 1\end{pmatrix}(v_1-\zeta v_2) = v_1-\zeta v_2 \quad \text{for every} \quad \beta \in K \ .$$

Hence $v_1 = \zeta v_2$. Thus up to a scalar multiple, there exists only one Bessel vector for ρ. We use this vector to define the <u>Bessel function</u> $J_\rho : G \to \mathbb{C}$ of ρ by

$$J_\rho(g) = [W_{v_1}\begin{pmatrix}1 & 0\\ 0 & 1\end{pmatrix}]^{-1} W_{v_1}(g).$$

Clearly $J_\rho(g)$ does not depend on the particular Bessel vector v_1 which is used in its definition. Note that J_ρ is also a Whittaker function for ρ. Therefore

$$J_\rho(gu) = J_\rho(ug) = \psi(u)J_\rho(g) \quad \text{for} \quad u \in U \quad \text{and} \quad g \in G \quad .$$

Also

$$(3) \qquad J_\rho(1) = 1 \quad \text{and} \quad J_\rho(a) = 0 \quad \text{if} \quad a \neq 1 \quad \text{and} \quad a \in A \quad .$$

Therefore, if a character ω of $K^\times$ is not exceptional for ρ, we have by Theorem 17.2 that

$$(4) \qquad \Gamma_\rho(\omega) = \sum_{x \in K^\times} J_\rho\begin{pmatrix}0 & 1\\ x & 0\end{pmatrix}\omega(x) \quad .$$

One can use this formula to define $\Gamma_\rho(\omega)$ also for the exceptional characters. We shall use this formula in the next two sections in order to compute Γ_ρ.

20. A computation of $\Gamma_\rho(\omega)$ for a non-cuspidal ρ.

Let ρ be a higher dimensional irreducible representation of G which is not cuspidal. Then ρ is a component of $\hat{\mu} = \mathrm{Ind}_B^G\mu$, where μ is a character of B that corresponds to the pair of characters (μ_1, μ_2) of $K^\times$. We may consider therefore V_ρ as an irreducible G-subspace of $\mathrm{Ind}_B^G V_\mu$. Every element of V_ρ appears then as a function $f: G \to \mathbb{C}$ such that

$$f(bg) = \mu(b)f(g) \qquad b \in B, \quad g \in G.$$

The action of G on V_ρ is given by $(\rho(s)f)(g) = f(gs)$.

We shall use this description of V_ρ in order to give a concrete Whittaker model for ρ. For every $f \in V_\rho$ let $W_f: G \to \mathbb{C}$ be the function defined by

(1) $$W_f(g) = \sum_{z \in K} f\left(w\begin{pmatrix} 1 & z \\ 0 & 1 \end{pmatrix} g\right)\psi(z)^{-1} \quad .$$

It is easy to check that W_f satisfies equalities (2) and (3) of section 16 characterizing Whittaker's functions

$$W_f(ug) = \psi(u)W_f(g) \qquad u \in U, \quad g \in G$$

$$W_{\rho(s)f}(g) = W_f(gs) \quad .$$

The map $f \mapsto W_f$ is therefore a linear map from V_ρ into $\mathrm{Ind}_U^G V_\psi$. We show that its kernel is zero by constructing a specific function $f_1 \in V_\rho$ such that $W_{f_1} \neq 0$. This implies that the map is injective and hence that $\{W_f | \ f \in V_\rho\}$ is a Whittaker model for ρ.

Using Bruhat's decomposition $G = B \cup BwU$, we define f_1 by

(2) $$f_1(b) = 0 \quad \text{and}$$

(3) $$f_1(bwu) = \mu(b)\psi(u) \quad \text{for } b \in B \text{ and } u \in U \quad .$$

Then f_1 is a non-zero element of $\mathrm{Ind}_U^G V_\psi$ and it satisfies

(4) $$f_1(gu) = \psi(u)f_1(g) \quad \text{for } g \in G \text{ and } u \in U \quad .$$

We prove that f_1 belongs to V_ρ.

If $\dim \rho = q+1$, then $V_\rho = \mathrm{Ind}_B^G V_\mu$, and there is nothing to prove. We can therefore assume that $\dim \rho = q$. In this case $\hat{\mu} = \rho' \oplus \rho$, where ρ' is a one-dimensional representation of G; hence $\mathrm{Ind}_B^G V_\mu = V_{\rho'} \oplus V_\rho$. We can therefore write $f_1 = f'+f$, where $f' \in V_{\rho'}$ and $f \in V_\rho$. Now, for a fixed $u \in U$ the functions $g \mapsto f'(gu)$ and $g \mapsto \psi(u)f'(g)$ belong to $V_{\rho'}$, whereas the functions $g \mapsto f(gu)$ and $g \mapsto \psi(u)f(g)$ belong to V_ρ. Using (4) we conclude that

(5) $$f'(gu) = \psi(u)f'(g) \quad \text{for } g \in G \quad .$$

By the proof of Lemma 8.6 the function f' satisfies

(6) $\quad f'(g) = f'(1)u_1(\det g) \quad$ for $g \in G$.

It follows from (5) that $f'(1) = \psi(u)f'(1)$ for every $u \in U$. Hence $f'(1) = 0$. This implies $f' = 0$ by (6). It follows that $f_1 = f \in V_\rho$. Computing W_{f_1} for $u' \in U$, we find

$$W_{f_1}(a') = \sum_{u\,U} f_1(wuu')\psi(u)^{-1} = \sum_{u\in U} \psi(uu')\psi(u)^{-1} = q\psi(u') .$$

Hence $W_{f_1} \neq 0$ and our contention is proved.

Note that (4) implies now that f_1 is a Bessel vector for ρ. We shall use this information in order to compute the Bessel function J_ρ, but first let us sum up the results proved in this section up to now.

<u>Lemma 20.1</u>: Let ρ be a non-cuspidal higher dimensional irreducible representation of G. Then the subspace $\{W_f \mid f \in V_\rho\}$ of $\mathrm{Ind}_U^G V_\psi$ defined by (1) is a Whittaker model for ρ, and the function f_1 defined by (2) and (3) is a Bessel vector for ρ in this model.

In order to compute $\Gamma_\rho(w)$, we now have to compute by (4) of section 19

(7) $$W_{f_1}\begin{pmatrix}0 & 1\\ x & 0\end{pmatrix} = \sum_{z\in K} f_1\left[\begin{pmatrix}0 & 1\\ 1 & 0\end{pmatrix}\begin{pmatrix}1 & z\\ 0 & 1\end{pmatrix}\begin{pmatrix}0 & 1\\ \alpha & 0\end{pmatrix}\right]\psi(z)^{-1} .$$

For $z = 0$ we have by (2) that

$$f_1\left[\begin{pmatrix}0 & 1\\ 1 & 0\end{pmatrix}\begin{pmatrix}0 & 1\\ x & 0\end{pmatrix}\right] = f_1\begin{pmatrix}x & 0\\ 0 & 1\end{pmatrix} = 0 .$$

If $z \neq 0$, then

$$\begin{pmatrix}0 & 1\\ 1 & 0\end{pmatrix}\begin{pmatrix}1 & z\\ 0 & 1\end{pmatrix}\begin{pmatrix}0 & 1\\ x & 0\end{pmatrix} = \begin{pmatrix}-z^{-1} & x\\ 0 & zx\end{pmatrix}\begin{pmatrix}0 & 1\\ 1 & 0\end{pmatrix}\begin{pmatrix}1 & (zx)^{-1}\\ 0 & 1\end{pmatrix};$$

hence by (3)

$$f_1\left[\begin{pmatrix}0 & 1\\ 1 & 0\end{pmatrix}\begin{pmatrix}1 & z\\ 0 & 1\end{pmatrix}\begin{pmatrix}0 & 1\\ x & 0\end{pmatrix}\right] = \mu_1(-z^{-1})\mu_2(zx)\psi(zx)^{-1} .$$

It follows from (7) that

$$W_{f_1}\begin{pmatrix} 0 & 1 \\ x & 0 \end{pmatrix} = \sum_{z\in K^\times} \mu_1(-z^{-1})\mu_2(zx)\psi((zx)^{-1}-z)$$

$$= \sum_{rs=-x^{-1}} \mu_1(r)^{-1}\mu_2(s)^{-1}\psi(s+r) \quad .$$

Also by (3)

$$W_{f_1}\begin{pmatrix} 1 & 0 \\ 0 & 1 \end{pmatrix} = \sum_{z\in K} f_1(w\begin{pmatrix} 1 & z \\ 0 & 1 \end{pmatrix})\psi(z)^{-1} = \sum_{z\in K} \psi(z)\psi(z)^{-1} = q \quad .$$

Hence

$$J_\rho\begin{pmatrix} 0 & 1 \\ x & 0 \end{pmatrix} = \frac{1}{q} \sum_{rs=-x^{-1}} \mu_1(r)^{-1}\mu_2(s)^{-1}\psi(r+s) \quad .$$

Substituting this value in formula (4) of section 19, we obtain

$$\Gamma_\rho(\omega) = \sum_{x\in K^\times} J_\rho\begin{pmatrix} 0 & 1 \\ x & 0 \end{pmatrix}\omega(x)$$

$$= \frac{1}{q} \sum_{x\in K^\times} \sum_{rs=-x^{-1}} \mu_1(r)^{-1}\mu_2(s)^{-1}\psi(r)\psi(s)\omega(x)$$

$$= \frac{1}{q} \sum_{x\in K^\times} \sum_{rs=x^{-1}} \mu_1(r)^{-1}\mu_2(s)^{-1}\psi(r)\psi(s)\omega(-r^{-1}s^{-1})$$

$$= \frac{\omega(-1)}{q} \sum_{r\in K^\times} \mu_1(r)^{-1}\omega(r)^{-1}\psi(r) \sum_{s\in K^\times}\mu_2(s)^{-1}\omega(s)^{-1}\psi(s) \quad .$$

Now recall that for a character χ of $K^\times$ one defines the Gauss sum $G(\chi,\psi)$ by

$$G(\chi,\psi) = \sum_{x\in K^\times} \chi(x)\psi(x) \quad .$$

We have therefore proved the following:

Theorem 20.2: Let μ_1,μ_2 be characters of $K^\times$ and let $\rho = \rho_{(\mu_1,\mu_2)}$ be the corresponding irreducible representation of G. If ω is a character of $K^\times$, then

$$\Gamma_\theta(\omega) = \frac{\omega(-1)}{q} G(\mu_1^{-1}\omega^{-1},\psi)G(\mu_2^{-1}\omega^{-1},\psi) \quad .$$

Remarks:

(1) It is well known that $|G(\chi,\psi)| = \sqrt{q}$ for every character χ of $K^\times$; hence $|\Gamma_\rho(\omega)| = 1$.

(2) If ψ' is another non-unit character of K^+, then there exists an $\alpha \in K^\times$ such that $\psi'(x) = \psi(\alpha x)$. It follows that $G(\chi,\psi') = \chi(\alpha)^{-1}G(\chi,\psi)$. Hence, if we denote by Γ'_ρ the Γ-function of ρ obtained by using ψ' instead of ψ, we obtain $\Gamma'_\rho(\omega) = \omega(\alpha)^2\mu_1(\alpha)\mu_2(\alpha)\Gamma_\rho(\omega)$.

21. A computation of $\Gamma_\rho(\omega)$ for a cuspidal ρ.

Let ν be a non-decomposable character of $L^\times$ and let $\rho = \rho_\nu$ be the corresponding cuspidal representation of G. Then $\rho(\omega')$ acts on a function $f: K^\times \to \mathbb{C}$ by

$$(1) \qquad (\rho(w')f)(x) = \sum_{\substack{y\in K^\times \\ Nt=u}} \nu(y)^{-1}j(yx)f(y) \quad ,$$

where $j = j_\nu$ is the function of $K^\times$ given by

$$j(u) = q^{-1} \sum_{\substack{t\in L^\times \\ Nt=u}} \psi(t+\bar{t})\nu(t)$$

(cf. Section 13). In particular, we apply (1) to the function $f'(x) = J_\rho\begin{pmatrix} x & 0 \\ 0 & 1\end{pmatrix}$ which is equal to 1 for $x = 1$ and equal to 0 if $x \neq 0$ (by (3) of Section 19). Also

$$(\rho(w')f')(x) = J_\rho\left[\begin{pmatrix} x & 0 \\ 0 & 1\end{pmatrix}\begin{pmatrix} 0 & 1 \\ -1 & 0\end{pmatrix}\right] = J_\rho\begin{pmatrix} 0 & x \\ -1 & 0\end{pmatrix}$$

Hence (1) implies

$$J_\rho\begin{pmatrix} 0 & x \\ -1 & 0\end{pmatrix} = j(x) \quad .$$

Also

$$\begin{pmatrix} 0 & 1 \\ x & 0 \end{pmatrix} = \begin{pmatrix} 0 & -x^{-1} \\ -1 & 0 \end{pmatrix}\begin{pmatrix} -x & 0 \\ 0 & -x \end{pmatrix} ;$$

hence by (2) of Section 13

$$J_\rho\begin{pmatrix} 0 & 1 \\ x & 0 \end{pmatrix} = (\rho\begin{pmatrix} -x & 0 \\ 0 & -x \end{pmatrix} J_\rho)\begin{pmatrix} 0 & -x^{-1} \\ -1 & 0 \end{pmatrix} = \nu(-x) J_\rho\begin{pmatrix} 0 & -x^{-1} \\ -1 & 0 \end{pmatrix} .$$

Hence, if ω is a character of $K^\times$, then

$$\Gamma_\rho(\omega) = \sum_{x\in K^\times} J_\rho\begin{pmatrix} 0 & 1 \\ x & 1 \end{pmatrix}\omega(x) = \sum_{x\in K^\times} \nu(-x)j(-x^{-1})\omega(x)$$

$$= -q^{-1} \sum_{x\in K^\times} \nu(-x)\omega(x) \sum_{t\bar{t}=-x^{-1}} \psi(t+\bar{t})\nu(t)$$

$$= q^{-1} \sum_{t\in L^\times} \nu(-t\bar{t})^{-1}\omega(t\bar{t})^{-1}\psi(t+\bar{t})\nu(t)$$

$$= q^{-1}\nu(-1) \sum_{t\in L^\times} \nu(\bar{t})\omega(t\bar{t})^{-1}\psi(t+\bar{t})$$

$$= q^{-1}\nu(-1) \sum_{t\in L^\times} \nu(t)\omega(t\bar{t})^{-1}\psi(t+\bar{t}) \quad .$$

Note that the last sum is a Gauss sum for the field L. Hence we have proved a result similar to Theorem 20.2.

Theorem 21.1: Let ν be a non-decomposable character of $L^\times$ and let $\rho = \rho_\nu$ be the corresponding cuspidal representation of G. Then

$$\Gamma_\rho(\omega) = \frac{\nu(-1)}{q} \sum_{t\in L^\times} \nu(t)\omega(t\bar{t})^{-1}\psi(t+\bar{t})$$

$$= \frac{\nu(-1)}{q} G_L(\nu\cdot(\omega\circ N_{L/K})^{-1} , \psi\circ Tr_{L/K})$$

for every character ω of K .

Remarks: As in the non-cuspidal case, $|\Gamma_\rho(\omega)| = 1$, since $|L| = q^2$. Also, if $\psi'(x) = \psi(\alpha x)$ is another character of $K^\times$, then $\Gamma'_\rho(\omega) = \nu(\alpha)^{-1}\omega(\alpha)^2\Gamma_\rho(\omega)$.

22. The characters of G.

We conclude our exposition on the representations of G with a computation of its characters table, i.e., the values of the irreducible higher dimensional characters at the representatives of the conjugacy classes of G. We recall that there are four families of representatives:

$$c_1(x) = \begin{pmatrix} x & 0 \\ 0 & x \end{pmatrix} \qquad x \in K^\times$$

$$c_2(x) = \begin{pmatrix} x & 1 \\ 0 & x \end{pmatrix} \qquad x \in K^\times$$

$$c_3(x,y) = \begin{pmatrix} x & 0 \\ 0 & y \end{pmatrix} \qquad x,y \in K^\times \text{ and } x \neq y$$

$$c_4(z) = \begin{pmatrix} 0 & -z\bar{z} \\ 1 & z+\bar{z} \end{pmatrix} \qquad z \in L-K \quad .$$

A. We start with the computation of the Gelfand-Graev character: Let $\hat{\psi} = \mathrm{Ind}_U^G \psi$. Denote by $\tilde{\psi}$ the function of G that coincides with ψ on U and is equal to zero elsewhere. Then $\chi_{\hat{\psi}}(g) = \sum_{s \in U\backslash G} \tilde{\psi}(sgs^{-1})$. In particular, if g is not conjugate to an element of U, then $\chi_{\hat{\psi}}(g) = 0$. The only eigenvalue of the elements of U is 1. It follows that the only representatives on which $\chi_{\hat{\psi}}$ must not vanish are $c_1(1)$ and $c_2(1)$. Clearly $\chi_{\hat{\psi}}(c_1(1)) = \chi_{\hat{\psi}}(1) = \dim \hat{\psi} = (q-1)q(q+1)$. In order to compute the value of $\chi_{\hat{\psi}}$ at $c_2(1) = \begin{pmatrix} 1 & 1 \\ 0 & 1 \end{pmatrix}$ we prove:

<u>Lemma 22.1</u>: $s\begin{pmatrix} 1 & 1 \\ 0 & 1 \end{pmatrix}s^{-1} \in U \iff s \in B$.

<u>Proof</u>: Let $s = \begin{pmatrix} \alpha & \beta \\ \gamma & \delta \end{pmatrix}$ be an element of G and let $s^{-1} = \begin{pmatrix} \alpha' & \beta' \\ \gamma' & \delta' \end{pmatrix}$. Then $s \in B$ if and only if $\gamma = 0$ (which is equivalent to $\gamma' = 0$). The lemma follows therefore from:

$$\begin{pmatrix}\alpha & \beta\\ \gamma & \delta\end{pmatrix}\begin{pmatrix}1 & 1\\ 0 & 1\end{pmatrix}\begin{pmatrix}\alpha' & \beta'\\ \gamma' & \delta'\end{pmatrix} = \begin{pmatrix}\alpha & \beta\\ \gamma & \delta\end{pmatrix}\left[\begin{pmatrix}1 & 0\\ 0 & 1\end{pmatrix} + \begin{pmatrix}0 & 1\\ 0 & 0\end{pmatrix}\right]\begin{pmatrix}\alpha' & \beta'\\ \gamma' & \delta'\end{pmatrix} =$$

$$= \begin{pmatrix}1 & 0\\ 0 & 1\end{pmatrix} + \begin{pmatrix}\alpha\gamma & \alpha\delta'\\ \gamma\gamma' & \gamma\delta'\end{pmatrix} = \begin{pmatrix}1+\alpha\gamma' & \alpha\delta'\\ \gamma\gamma' & 1+\gamma\delta'\end{pmatrix}. \qquad //$$

The lemma implies that

$$\chi_{\hat\psi}\begin{pmatrix}1 & 1\\ 0 & 1\end{pmatrix} = \sum_{s\in U\backslash B} \psi\left(s\begin{pmatrix}1 & 1\\ 0 & 1\end{pmatrix}s^{-1}\right) =$$

$$= \sum_{\alpha_1\delta\in K^\times} \psi\left[\begin{pmatrix}\alpha & 0\\ 0 & \delta\end{pmatrix}\begin{pmatrix}1 & 1\\ 0 & 1\end{pmatrix}\begin{pmatrix}\alpha^{-1} & 0\\ 0 & \delta^{-1}\end{pmatrix}\right] = \sum_{\alpha,\delta\in K^\times} \psi\begin{pmatrix}1 & \alpha\delta^{-1}\\ 0 & 1\end{pmatrix} =$$

$$= (q-1) \sum_{x\in K^\times} \psi(x) = 1-q .$$

We have therefore proved:

<u>Theorem 22.2</u>: Let $\hat\psi = \mathrm{Ind}_U^G\psi$. Then

$$\chi_{\hat\psi}(c_1(1)) = (q-1)q(q+1) ,$$

$$\chi_{\hat\psi}(c_1(x)) = 0 \quad \text{if } x \neq 1 ,$$

$$\chi_{\hat\psi}(c_2(1)) = 1-q ,$$

$$\chi_{\hat\psi}(c_2(x)) = 0 \quad \text{if } x \neq 1 ,$$

$$\chi_{\hat\psi}(c_3(x,y)) = 0 ,$$

$$\chi_{\hat\psi}(c_4(z)) = 0 .$$

B. Let μ be a character of B which is defined by the pair of characters (μ_1,μ_2) of $K^\times$ through the formula

$$\mu\begin{pmatrix}\alpha & \beta\\ 0 & \delta\end{pmatrix} = \mu_1(\alpha)\mu_2(\delta) .$$

Let $\hat\mu = \mathrm{Ind}_B^G\mu$ and compute $\chi_{\hat\mu}$. First

$$\chi_{\hat\mu}(c_1(x)) = \chi_{\hat\mu}\begin{pmatrix} x & 0 \\ 0 & x\end{pmatrix} = \sum_{s\in B\backslash G} \mu\left(s\begin{pmatrix} x & 0 \\ 0 & x\end{pmatrix}s^{-1}\right) = (G:B)\mu\begin{pmatrix} x & 0 \\ 0 & x\end{pmatrix} =$$

$$= (q+1)\mu_1(x)\mu_2(x) \quad .$$

Second, as in Lemma 22.1, one proves that

$$s\begin{pmatrix} x & 1 \\ 0 & x\end{pmatrix}s^{-1} \in B \iff s \in B \quad .$$

Hence

$$\chi_{\hat\mu}(c_2(x)) = \mu\begin{pmatrix} x & 1 \\ 0 & x\end{pmatrix} = \mu_1(x)\mu_2(x) \quad .$$

In order to compute the value of $\chi_{\hat\mu}$ at $c_3(x,y)$, we note that a direct computation shows that

(1) $$\begin{pmatrix} \alpha & \beta \\ \gamma & \delta\end{pmatrix}\begin{pmatrix} x & 0 \\ 0 & y\end{pmatrix}\begin{pmatrix} \alpha & \beta \\ \gamma & \delta\end{pmatrix}^{-1} \in B \iff \gamma = 0 \text{ or } \delta = 0.$$

Elements of $B\backslash G$ that satisfy the left-hand side of (1) are therefore 1 and w. Indeed,

$$\begin{pmatrix} \alpha & \beta \\ \gamma & 0\end{pmatrix} = \begin{pmatrix} \beta & \alpha \\ 0 & \gamma\end{pmatrix}\begin{pmatrix} 0 & 1 \\ 1 & 0\end{pmatrix} .$$

Hence

$$\chi_{\hat\mu}(c_3(x,y)) = \mu\begin{pmatrix} x & 0 \\ 0 & y\end{pmatrix} + \mu\left[\begin{pmatrix} 0 & 1 \\ 1 & 0\end{pmatrix}\begin{pmatrix} x & 0 \\ 0 & y\end{pmatrix}\begin{pmatrix} 0 & 1 \\ 1 & 0\end{pmatrix}\right] =$$

$$= \mu\begin{pmatrix} x & 0 \\ 0 & y\end{pmatrix} + \mu\begin{pmatrix} y & 0 \\ 0 & x\end{pmatrix} = \mu_1(x)\mu_2(y) + \mu_1(y)\mu_2(x) \quad .$$

Finally recall that the eigenvalues of the elements $c_4(z)$ do not belong to K. Hence the $c_4(z)$ are not conjugate to elements of B and therefore $\chi_{\hat\mu}(c_4(z)) = 0$.

B1. If $\mu_1 \neq \mu_2$, then $\hat\mu$ is an irreducible representation of G. Its character has therefore been computed.

B2. If $\mu_1 = \mu_2$, then $\hat\mu = \rho'\oplus\rho$, where ρ' is a one-dimensional character given by the formula $\rho'(g) = \mu_1(\det g)$ (by Lemma 8.6). Hence $\rho'(c_1(x)) = \mu_1(x)^2$, $\rho'(c_2(x)) = \mu_1(x)^2$, $\rho'(c_3(x,y)) = \mu_1(x)\mu_1(g)$ and $\rho'(c_4(z)) = z\bar{z}$. The values of χ_ρ can therefore be computed from the above data and from the formula $\chi_\rho(g) = \chi_{\hat{\mu}}(g)-\mu_1(\det g)$.

C. The most difficult computation is presented by cuspidal representations. Let ν be a non-decomposable character of $L^\times$ and let $\rho = \rho_\nu$ be the corresponding cuspidal representation. As V_ρ we choose, as we did in Section 13, the space $F(K^\times,\mathbb{C})$ of all functions $f\colon K^\times \to \mathbb{C}$. Let W_f be the Whittaker functions for ρ and let J_ρ be its Bessel function. Then

$$(2) \qquad W_f(g) = \sum_{y\in K^\times} W_f\begin{pmatrix} y & 0 \\ 0 & 1 \end{pmatrix} J_\rho\left(g\begin{pmatrix} y^{-1} & 0 \\ 0 & 1 \end{pmatrix}\right) .$$

Indeed, both sides of (2) are Whittaker functions and they coincide on A, since $J_\rho(a) = 0$ if $1 \neq a \in A$ and $J_\rho(1) = 1$. Hence by Lemma 18.1 they coincide on g's (recall that $J(V_\rho) = 0$). The same lemma implies that for every function $\varphi\colon K^\times \to \mathbb{C}$, there exists an $f \in V_\rho$ such that

$$W_f\begin{pmatrix} x & 0 \\ 0 & 1 \end{pmatrix} = \varphi(x) \quad \text{for} \quad x \in K^\times .$$

It follows by (2) that

$$(3) \qquad (\rho(g)\varphi)(x) = W_{\rho(g)f}\begin{pmatrix} x & 0 \\ 0 & 1 \end{pmatrix} = W_f\left[\begin{pmatrix} x & 0 \\ 0 & 1 \end{pmatrix} g\right]$$

$$\sum_{y\in K^\times} \varphi(y) J_\rho\left[\begin{pmatrix} x & 0 \\ 0 & 1 \end{pmatrix} g \begin{pmatrix} y^{-1} & 0 \\ 0 & 1 \end{pmatrix}\right] .$$

The space $F(K^\times,\mathbb{C})$ has a natural basis $\{\varphi_\alpha \mid \alpha \in K^\times\}$, where $\varphi_\alpha(\alpha) = 1$ and $\varphi_\alpha(x) = 0$ if $x \neq \alpha$. Substituting $\varphi = \varphi_\alpha$ in (3), we have

$$(4) \qquad (\rho(g)\varphi_\alpha)(x) = J_\rho\left[\begin{pmatrix} x & 0 \\ 0 & 1 \end{pmatrix} g \begin{pmatrix} \alpha^{-1} & 0 \\ 0 & 1 \end{pmatrix}\right] .$$

The collection of all the elements on the right-hand side of (4) gives a $(q-1)\times(q-1)$ matrix representing $\rho(g)$. The trace of this matrix is

$$(5) \qquad \chi_\rho(g) = \sum_{x\in K^\times} J_\rho\left[\begin{pmatrix} x & 0 \\ 0 & 1\end{pmatrix} g \begin{pmatrix} x^{-1} & 0 \\ 0 & 1\end{pmatrix}\right] .$$

In order to apply this formula we compute J_ρ for some values of G. Let f_1 be a Bessel vector for ρ. Then

$$\rho\begin{pmatrix} x & \alpha \\ 0 & x\end{pmatrix} f_1 = \nu(x)\psi(x^{-1}\alpha) f_1$$

by (3) of Section 13. Hence

$$(6) \qquad J_\rho\left[g\begin{pmatrix} x & \alpha \\ 0 & x\end{pmatrix}\right] = W_{f_1}(1)^{-1} W_{f_1}\left[\begin{pmatrix} x & \alpha \\ 0 & x\end{pmatrix}\right] =$$

$$= W_{f_1}(1)^{-1} W_\rho\begin{pmatrix} x & \alpha \\ 0 & x\end{pmatrix} f_1(g) = \nu(x)\psi(x^{-1}\alpha) J(g) \quad .$$

The computation of χ_ρ for $c_1(x)$ is now

$$\chi_\rho\begin{pmatrix} x & 0 \\ 0 & x\end{pmatrix} = \sum_{\alpha\in K^\times} J_\rho\left[\begin{pmatrix} \alpha & 0 \\ 0 & 1\end{pmatrix}\begin{pmatrix} x & 0 \\ 0 & x\end{pmatrix}\begin{pmatrix} \alpha^{-1} & 0 \\ 0 & 1\end{pmatrix}\right] = (q-1) J_\rho\begin{pmatrix} x & 0 \\ 0 & x\end{pmatrix} = (q-1)\nu(x) .$$

For $c_2(x)$ we have:

$$\chi_\rho\begin{pmatrix} x & 1 \\ 0 & x\end{pmatrix} = \sum_{\alpha\in K^\times} J_\rho\left[\begin{pmatrix} \alpha & 0 \\ 0 & 1\end{pmatrix}\begin{pmatrix} x & 1 \\ 0 & x\end{pmatrix}\begin{pmatrix} \alpha^{-1} & 0 \\ 0 & 1\end{pmatrix}\right] = \sum_{\alpha\in K^\times} J_\rho\begin{pmatrix} x & \alpha \\ 0 & x\end{pmatrix}$$

$$= \nu(x) \sum_{\alpha\in K^\times} \psi(x^{-1}\alpha) = -\nu(x) \quad .$$

For $c_3(x,y)$ we have:

$$\chi_\rho\begin{pmatrix} x & 0 \\ 0 & y\end{pmatrix} = \sum_{\alpha\in K^\times} J_\rho\left[\begin{pmatrix} \alpha & 0 \\ 0 & 1\end{pmatrix}\begin{pmatrix} x & 0 \\ 0 & y\end{pmatrix}\begin{pmatrix} \alpha^{-1} & 0 \\ 0 & 1\end{pmatrix}\right] = (q-1) J_\rho\begin{pmatrix} x & 0 \\ 0 & y\end{pmatrix} =$$

$$= (q-1) J_\rho\left[\begin{pmatrix} xy^{-1} & 0 \\ 0 & 1\end{pmatrix}\begin{pmatrix} y & 0 \\ 0 & y\end{pmatrix}\right] = (q-1)\nu(g) J_\rho\begin{pmatrix} xy^{-1} & 0 \\ 0 & 1\end{pmatrix} = 0 ,$$

since $x \neq y$.

The computation of $\chi_\rho(c_4(z))$ does not rely on formula (5), but rather on formulas (15) and (16) of Section 13. Formula (15) of Section 13 implies that $(k(y,x;g))_{x,y\in K^\times}$ is the matrix representing $\rho(g)$ with respect to the natural basis. Its trace is

$$\chi_\rho(g) = \sum_{x\in K^\times} k(x,x;g) \quad .$$

In particular let $a = z+\bar{z}$ and $b = -z\bar{z}$. Then

$$\chi_\rho(c_4(zz) = \sum_{x\in K^\times} k(x,x;\begin{pmatrix}0 & b\\ 1 & a\end{pmatrix}) =$$

$$= -q^{-1} \sum_{x\in K^\times} \psi(ax) \sum_{\substack{u\bar{u}=-b\\ u\in L^\times}} \psi(-x(u+\bar{u}))\nu(u)$$

$$= -q^{-1} \sum_{\substack{u\bar{u}=-b\\ u\in L^\times}} \nu(u) \sum_{x\in K^\times} \psi(x(a-(u+\bar{u}))) \quad .$$

If $u = z$ or $u = \bar{z}$, then $a = u+\bar{u}$; hence $\sum_{x\in K^\times} (x(a-(u+\bar{u}))) = q-1$. If $u \neq z$ and $b \neq z$ and $u\bar{u} = -b$, then $a \neq u+\bar{u}$; hence $\sum_{x\in K^\times}\psi(x(a-(u+\bar{u}))) = -1$. It follows that

$$\chi_\rho(c_4(z)) = -q^{-1}\left[(q-1)\nu(z)+(q-1)\nu(\bar{z}) - \sum_{\substack{u\neq z,\bar{z}\\ u\bar{u}=-b}} \nu(u)\right] =$$

$$= -q^{-1}\left[(q\nu(z)+q\nu(\bar{z}) - \sum_{\substack{u\in L\\ u\bar{u}=-b}} \nu(u)\right] = -\nu(z)-\nu(\bar{z}) \quad ,$$

since by Lemma 12.4 ν is non-decomposable and hence $\sum_{u\bar{u}=-b} \nu(u) = 0.$

We sum up the character values computed in this section in the following table. Here ν is a non-decomposable character of $L^\times$, μ_1,μ_2 are characters of $K^\times$ and $\delta_{1,x}$ is the Kronecker symbol, equal to 1 if $x = 1$ and to zero if $1 \neq x$. The third column gives the number of the representations of the corresponding type. The fourth, fifth, sixth and seventh columns give the values of the characters of the corresponding representations.

Repr.	Dimension	Number	$c_1(x)$	$c_2(x)$	$c_3(x,y)$	$c_4(z)$
ρ_ν	$q-1$	$\frac{1}{2}(q^2-q)$	$(q-1)\nu(x)$	$-\nu(x)$	0	$-\nu(z)-\nu(\bar{z})$
$\rho(\mu_1,\mu_2)$	q	$q-1$	$q\mu_1(x)^2$	0	$\mu_1(xy)$	$-\mu_1(z\bar{z})$
$\rho(\mu_1,\mu_2)$	$q+1$	$\frac{1}{2}(q-1)(q-2)$	$(q+1)\mu_1(x)\mu_2(x)$	$\mu_1(x)\mu_2(x)$	$\mu_1(x)\mu_2(y)+\mu_1(y)\mu_2(x)$	0
$\mathrm{Ind}_U^G\psi$	$(q-1)q(q+1)$	1	$(q-1)q(q+1)\delta_{1,x}$	$(1-q)\delta_{1,x}$	0	0

REFERENCES

1. S. Lang, Algebra, Addison-Wesley, Reading, Mass., 1955.

2. J.P. Serre, Représentation linéaires des groupes finis, Hermann, Paris, 1971.

INDEX

ABCDEFGHIJ–AMS–89876543